Building the Anti-Racist University

In the new arena for anti-racist work in which we find ourselves, the neoliberal, 'post-race' university, this interdisciplinary collection demonstrates common global political concerns about racism in Higher Education. It highlights a range of issues regarding students, academic staff and knowledge systems, and all of the contributions seek to challenge the complacency of the 'post-race' present that is dominant in North-West Europe and North America, Brazil's mythical 'racial democracy' and South Africa's post-apartheid 'rainbow nation'.

The collection makes clear that we are not yet past the need for anti-racist institutional action because of the continuing impact of coloniality on and in these nations. From within the colonial psyche which still exists in the 21st century these nations actively deracinate politics, subjectivities, political economy and affective relationalities when they re-imagine themselves to be 'post-race' states where all citizens can have a share in the good life because now only class matters. Universities have also taken on the mantle of upholding societal 'post-race' status through ineffective equality and diversity policies and strategies.

The collection makes the case for the urgent need to decolonize the university in 'post-race', neoliberal times through a focus on institutional racism in HEIs in Canada, Brazil, South Africa, the UK and the USA. As such it addresses institutional whiteness; the transformation of organizational cultures; the presence and experiences of Black people, People of Colour and Indigenous people in HEIs; the development of curriculum interventions; widening participation and organizational change; and future directions for racial equality and diversity in a 'post-race' era.

This book was originally published as a special issue of *Race Ethnicity and Education*.

Shirley Anne Tate is Honorary Professor, Chair for Critical Studies in Higher Education Transformation, Nelson Mandela University, South Africa. Her area of research is Black diaspora studies broadly. Her work focuses on the intersections of 'race' and gender, and her research interests are institutional racism, the body, affect, 'mixed race', beauty, 'race' performativity and Caribbean decolonial studies.

Paul Bagguley is Reader in Sociology in the School of Sociology and Social Policy at the University of Leeds, UK. His main interests are in the sociology of social movements, racism and ethnicity. In the field of racism and ethnicity studies he has worked on the 2001 riots, South Asian women and higher education, the impacts of the 7/7 London bombings on different ethnic and religious groups in West Yorkshire and the cosmopolitanism of traditional British retail markets. More broadly his research interests and publications have encompassed economic sociology, urban studies and social theory.

Building the Anti-Racist University

Edited by
Shirley Anne Tate and Paul Bagguley

Routledge
Taylor & Francis Group

LONDON AND NEW YORK

First published 2019
by Routledge
2 Park Square, Milton Park, Abingdon, Oxon, OX14 4RN, UK

and by Routledge
711 Third Avenue, New York, NY 10017, USA

Routledge is an imprint of the Taylor & Francis Group, an informa business

British Library Cataloguing-in-Publication Data
A catalogue record for this book is available from the British Library

ISBN13: 978-0-367-00151-3

Typeset in Minion Pro
by codeMantra

Publisher's Note
The publisher accepts responsibility for any inconsistencies that may have arisen during the conversion of this book from journal articles to book chapters, namely the possible inclusion of journal terminology.

Disclaimer
Every effort has been made to contact copyright holders for their permission to reprint material in this book. The publishers would be grateful to hear from any copyright holder who is not here acknowledged and will undertake to rectify any errors or omissions in future editions of this book.

Contents

CONTENTS

Citation Information

The chapters in this book were originally published in *Race Ethnicity and Education*, volume 20, issue 3 (May 2017). When citing this material, please use the original page numbering for each article, as follows:

Chapter 6

Affirmative action in Brazil and building an anti-racist university
Joaze Bernardino-Costa and Ana Elisa De Carli Blackman
Race Ethnicity and Education, volume 20, issue 3 (May 2017) pp. 372–384

Chapter 7

The challenge of creating a more diverse economics: lessons from the UCR minority pipeline project
Gary A. Dymski
Race Ethnicity and Education, volume 20, issue 3 (May 2017) pp. 385–400

Chapter 8

Dealing with difficult conversations: anti-racism in youth & community work training
Diana Watt
Race Ethnicity and Education, volume 20, issue 3 (May 2017) pp. 401–413

Chapter 9

From Liverpool to New York City: behind the veil of a Black British male scholar inside higher education
Mark Christian
Race Ethnicity and Education, volume 20, issue 3 (May 2017) pp. 414–428

For any permission-related enquiries please visit:
http://www.tandfonline.com/page/help/permissions

Notes on Contributors

Ryan P. Adserias is affiliated with Wisconsin's Equity and Inclusion Laboratory at the University of Wisconsin-Madison, USA.

Paul Bagguley is Reader in Sociology in the School of Sociology and Social Policy at the University of Leeds, UK. His main interests are in the sociology of social movements, racism and ethnicity. In the field of racism and ethnicity studies he has worked on the 2001 riots, South Asian women and higher education, the impacts of the 7/7 London bombings on different ethnic and religious groups in West Yorkshire and the cosmopolitanism of traditional British retail markets. More broadly his research interests and publications have encompassed economic sociology, urban studies and social theory.

Joaze Bernardino-Costa is Professor of Sociology in the Department of Sociology at the University of Brasília, Brazil.

Vivienne Bozalek is Director of Teaching and Learning at the University of the Western Cape, Cape Town, South Africa

Ronelle Carolissen is Professor of Community Psychology and Vice Dean of Teaching & Learning in the Faculty of Education at Stellenbosch University, South Africa.

LaVar J. Charleston is Assistant Vice Chancellor for Student Diversity, Engagement and Success at the University of Wisconsin-Whitewater, USA.

Mark Christian is full Professor and Chair of Africana Studies at Lehman College, City University of New York, USA.

Ana Elisa De Carli Blackman is a faculty member in the Department of Sociology at the University of Brasília, Brazil.

Enakshi Dua is Professor of Women's Studies in the Department of Women's Studies at York University, Toronto, Canada.

Gary A. Dymski is Professor and Chair in Applied Economics at Leeds University Business School, UK.

Pete Harris is Senior Lecturer and Course Coordinator for Criminology in the Department of Youth and Community Work at Newman University, Birmingham, UK.

Chris Haywood is Reader in Critical Masculinity Studies in the Department of Media and Cultural Studies at Newcastle University, Newcastle upon Tyne, UK.

NOTES ON CONTRIBUTORS

Frances Henry is Professor Emerita in the Department of Anthropology at York University, Toronto, Canada.

Jerlando F. L. Jackson is Director and Chief Research Scientist at Wisconsin's Equity and Inclusion Laboratory at the University of Wisconsin-Madison, USA.

Carl James is Professor and Jean Augustine Chair in Education, Community & Diaspora in the Faculty of Education and the Department of Sociology at York University, Toronto, Canada.

Audrey Kobayashi is Professor in the Department of Geography and Planning at Queens University, Kingston, Canada.

Ian Law is Research Professor in the School of Social Sciences at Södertörn University, Stockholm, Sweden.

Peter Li is Professor Emeritus in the Department of Sociology at the University of Saskatchewan, Saskatoon, Canada.

Mairtin Mac an Ghaill is Interim Director of Graduate School in the Department of Education at Newman University, Birmingham, UK.

Howard Ramos is Professor and Associate Dean of Research in the Faculty of Arts and Social Sciences in the Department of Sociology at Dalhousie University, Halifax, Canada.

Malinda S. Smith is Professor in the Department of Political Science in the Faculty of Arts at the University of Alberta, Edmonton, Canada.

Shirley Anne Tate is Honorary Professor, Chair for Critical Studies in Higher Education Transformation, Nelson Mandela University, South Africa. Her area of research is Black diaspora studies broadly. Her work focuses on the intersections of 'race' and gender, and her research interests are institutional racism, the body, affect, 'mixed race', beauty, 'race' performativity and Caribbean decolonial studies.

Diana Watt is Senior Lecturer in Childhood, Youth and Education Studies at Manchester Metropolitan University, UK.

Building the anti-racist university: next steps

Shirley Anne Tate and Paul Bagguley

Introduction: first steps

This special issue emerged out of the continuing concern with how best to deal with institutional racism in higher education institutions (HEIs) that we have long shared as colleagues in the Center for Ethnicity and Racism Studies (CERS) at the University of Leeds, as discussed by Ian Law in this volume. The 2013 conference 'Building the Anti-racist University: Next Steps' was focused on looking forward to what needed to be done now in the twenty-first century drawing on twentieth/twenty-first century experience of institutional gains followed by their attrition in some cases and fundamental institutional inertia in others. Both of these responses to addressing institutional racism worked against organizational change even as equality and diversity policies aimed at changing the face of universities were instituted.

The papers in this special issue are the results of the thinking instantiated by the call for papers and the transdisciplinary, transnational theoretical, and practice-based discussions at the conference on experiences of institutional racial equality change processes and strategies as both partial successes and abject failures. We take both successes and failures forward as lessons learned into the new arena for anti-racist work in which we find ourselves, the neoliberal, 'post-race' university which by and large still caters for national/international elites, where some knowledge is commodified on a global scale and others continue to be erased. What is distinctive about this special issue is the international character of the collection demonstrating common political concerns globally about racism in higher education. Yet there remain some puzzling absences – no contribution from mainland Europe, the Caribbean, or Australia and New Zealand for example. This may perhaps reflect our networks, how we framed the conference, or be an indication that racism in higher education does not get much attention in these contexts in which anti-black and anti-Indigenous racisms persist. Notwithstanding these absences, one goal of this special issue is to further expand the global debate on racism and anti-racism in universities. The papers highlight a multiple range of issues regarding students, academic staff, and knowledge systems but all seek to challenge the complacency of the 'post-race' present that is dominant in, northwest Europe and North America, Brazil's mythical 'racial democracy' and South Africa's post-apartheid 'rainbow nation'. The papers also originate from a variety of disciplines – Sociology, Economics, Pyschology, Education, and Youth and Community Work.

For the countries represented by the papers – Brazil, South Africa, Canada, the USA, and the UK – what is clear is that we are not yet past the need for anti-racist institutional action. What these nations share in common is that they were all touched by the machinations of European empire whether as colonized or colonizer. This has led to the instantiation of European whiteness as superior and abjection of the difference of racialized others. From within this colonial psyche which still exists in the twenty-first century, these nations actively deracinate politics, subjectivities, political economy, and affective relationalities when they re-imagine themselves to be 'post-race' states where all citizens can have a share in the good life because now only class matters. Universities have also taken on the mantle of upholding societal 'post-race' status through those very same equality and diversity policies and strategies which have not been effective (Ahmed 2012). Frances Henry et al's article on higher education in Canada foregrounds *racism* as a critical variable shaping racialized and Indigenous peoples' lives and experiences.

This issue is pronounced in Canadian universities, where employment equity, diversity, and other policies aimed at equality amount to no more than well-worded mission statements and some minor cosmetic changes which leave structural racial inequality intact. In Canada inequality, indifference, and reliance on outmoded conservative traditions characterize the modern neoliberal university which continues to work on racial lines. Whether one examines representation in terms of numbers of racialized and Indigenous faculty members and their positioning within the system, their earned income as compared to white faculty, their daily life experiences of racism within the university as workplace irrespective of status, or interactions with colleagues and students, the results are that racialized and Indigenous faculty and the disciplines or areas of their expertise are, on the whole, low in numbers and even lower in terms of power, prestige, and influence within the HEIs. From the viewpoint of the USA, Ryan P. Adserias, Lavar J. Charleston, and Jerlando F.L. Jackson assert that implementing racial diversity agendas within decentralized, loosely coupled, and change-resistant institutions such as colleges and universities is a global challenge. They see a shift in organizational culture as imperative in order to produce the change needed for a diversity agenda to thrive. This article synthesizes the literature on proven strategies and offers case studies of how a variety of leadership styles has and can fuel much needed racial diversity efforts or lead to institutional inertia.

More work needs to be done into the twenty-first century because of, not in spite of, the 'post-race' consensus in order to develop a maximal, transformative approach to institutional change, rather than a minimal meeting of legal obligations in those countries where an anti-discrimination framework exists. In the UK, progress in the field of anti-racism in HEIs has slowed and has dissipated across the sector within a proliferation of policy statements on equity, diversity, and harassment as well as ethnic monitoring of staff and student access and progression, for example. These approaches have been inadequate and do not reflect the necessary institutional effort required to establish real and lasting anti-racism in the UK higher education sector, or indeed, in Canada, the USA, Brazil, and South Africa built on a foundation of innovative and effective policy and practice. This special issue draws together the foci emerging from the debates within each paper on curriculum, pedagogy, access, policy, process, experience, outcomes, racialization, and racism in HEIs in Canada, the USA, the UK, Brazil, and South Africa to help in crafting an agenda for building the global anti-racist university into the 'post-race' twenty-first century.

To aid in this endeavor, the papers in this special issue look at the following key themes in their locally contextualized debates and research on institutional racism in HEIs in Canada, Brazil, South Africa, the UK, and USA:

(1) Institutional whiteness: How is it produced and reproduced through affect, structures, and processes? How might it be resisted and transformed?

(2) Transforming organizational cultures: What are the challenges of such transformation? What are the conflicts and contradictions of transforming HEIs 'from within'? Are our efforts always destined to be turned into another managerial process? What role does intersectionality play in transforming organizational cultures?

(3) The black and minority ethnic (BME) and Indigenous presence and experience in HEIs: how can we best map these in terms of both staff and students? Can we draw in meaningful ways on these experiences to produce change in HEIs' approaches to curriculum, pedagogy, recruitment, retention, and progression?

(4) Developing curriculum interventions: what can be done to enable anti-racism within a context of professional autonomy, disciplinary inertia, and organizational resistance?

(5) Widening participation and organizational change: What does widening participation mean in the context of anti-racism? Should anti-racism be a part of the outcomes of higher education curricula?

(6)   Future directions for racial equality and diversity in a 'post-race' era; what are the implications and symptoms of 'post-race' for HEIs? What impact does 'post-race' have on the possibility for the development of anti-racist strategies?

Institutional whiteness is shared across all of the papers in the issue so let us turn next to briefly look at whiteness and institutional racism in contemporary university spaces in the 'post-race' UK.

Whiteness, institutional racism and universities as 'post-race' spaces

We began the debate within Racism Studies about whether or not we are yet 'post-race' societies some time ago (Goldberg 2015). Whatever side of the debate on which we fall what this special insists is that institutional racism is still very much a part of the fabric of the university spaces we inhabit, texturing our experiences and this is the case no matter how much we might wish that it were otherwise. Academia is an institution in which faculty and administration continue to be predominantly white especially at professor, vice chancellor, and top management levels and the curricula continue to be unashamedly white as well. Continuing dissatisfaction with this state of affairs led to the emergence of student-led campaigns in the UK on 'why is my curriculum white?' (http://www.nus.org.uk/en/news/why-is-my-curriculum-white/) and 'why isn't my professor black?' (http://www.dtmh.ucl.ac.uk/isnt-professor-black-reflection/). These concerns with the lack of change in terms of racial justice transformation have led over the last few years to the mobilization of thousands of students to public meetings in universities across the country and their political attachment to other global campaigns such as '#Black Lives Matter' in the USA and 'Rhodes Must fall' in South Africa. Much of this public debate and campus-based campaigning has emerged since our conference, yet they indicate its political timeliness.

The UK student mobilizations became more apparent after a historic panel at the UCL on 10 March 2014 entitled 'Why isn't my professor black?' The members of the panel were Professor Michael Arthur President and Provost (UCL), Dr Deborah Gabriel (Founder and CEO of Black British Academics), Dr Lisa Palmer (Newman University), Dr Shirley Anne Tate (University of Leeds), Dr William Ackah (Birkbeck College, University of London), and Dr Nathaniel Adam Tobias Coleman (UCL) who organized the panel. The event was 'sold out' within days and a bigger venue had to be arranged in order to seat the hundreds of people who attended. The UCL panel is widely seen to have been the catalyst for anti-racist student campaigns and student calls to decolonize the university in the UK. At this panel, the vice chancellor of UCL asserted that that university would develop the first Black Studies program in the UK to show its commitment to this area of academic endeavor globally. However, this has not yet materialized at UCL and progress on this achievement seems to have dissipated. In South Africa and the UK, there has also been the 'Rhodes Must Fall' campaign and in South Africa the 'Fees Must Fall' campaign. All of these student-led mobilizations have been a call to action for anti-racist change not just within universities but also societally. Cynically, UK universities have responded with once yearly well-publicized Black History Month events as part of their equality and diversity strategies, part of a public demonstration of their commitment to anti-racist change. These are sometimes run as public events by their Public Relations offices to show 'there is no racism here', irrespective of the fact of the shameful BME employment statistics within UK HEIs at present and the prevailing issue of BME student lack of achievement. There continues to be under-representation of BME staff even while there has been a year-on-year increase in BME students (Equality Challenge Unit 2015 http://www.ecu.ac.uk/publications/equality-higher-education-statistical-report-2015/ accessed 1 August, 2016). The numbers of BME staff have increased from 4.8% in 2003/4 to 6.7% in 2013/14 (ECU 2015). Further, black staff members continue to be low paid and low status in comparison with white colleagues (ECU 2015).

Whilst much previous work in the UK focused on racial inequalities in access to university (McManus et al. 1998; Connor et al. 2004; Bagguley and Hussain 2007) more recent work has revealed a significant 'attainment gap' between white, black, and ethnic minority students. Data from the United Kingdom's Equality Challenge Unit (http://www.ecu.ac.uk/guidance-re-sources/student-recruitment-retention-attainment/student-attainment/degree-attainment-gaps/ accessed 1 August 2016) showed that in 2012/13 57.1% of UK-domiciled BME students received an upper second class or first class degree, compared with 73.2% of white British students. This is what the ECU refers to as an attainment gap of 16.1%. Whilst the gap varies between minority ethnic groups, 43.8% of self-classified 'Black Other' students achieved a higher class of degree – a gap of 29.4% compared to white students. Such an attainment gap should make universities ponder what it is about, what happens within their walls, classrooms, and curricula that suppresses the emergence of BME student excellence. Students have already highlighted those aspects of university life which impact their experiences negatively in terms of the campaigns mentioned above, that is, continuing institutional racism, curricula which continue to be Euro-centric and faculty which do not reflect the UK's demographic diversity. These very issues were raised in terms of schooling by Bernard Coard's (1971) *How the West Indian Child is Made Educationally Subnormal in British Schools* and Maureen Stone's (1981) *The Education of the Black Child in Britain: The Myth of Multi-racial Education*. One could say then that the UK education system has not moved past race and, indeed, is configured to maintain the dominance of those racialized as white.

This dominance is also maintained through a second feature of the university landscape in the UK that has been receiving increasing attention. That is, the lack of progression of black and ethnic minority students into the academic workforce. For example, at the time of writing there were only 18 black women full professors in the UK (the Times Higher 17 August 2016). One particular paradox here is that whilst black and ethnic minority students are more likely than white students to study for a taught Master's, they are less likely to move on to a PhD (http://www.hefce.ac.uk/pubs/year/2016/201614/) which is the first step towards an academic career in the UK. In contrast, white graduates were almost twice as likely as BME graduates to go on to a research degree soon after graduating. This research by the government's Higher Education Funding Council demonstrates some level of official concern, but this contrasts with the lack of real action for change within universities, such as student mentoring and scholarship possibilities. Indeed, if the majority of UK BME students attend non-Russell Group universities this already means that they stand less chance of getting an ESRC/AHRC scholarship than their Russell Group counterparts. The organization of scholarship funding through the doctoral training centers/partnerships model potentially could be the location of unwitting racial exclusion even though on the face of it the system seems to be operating on a meritocratic basis. Such enduring inequalities at the heart of UK HEIs supposedly built upon those long-held Eurocentric virtues of fairness and meritocracy reveal an ongoing monumental structural racial inequality and ongoing racist practices.

'Post-race' we are not (Goldberg 2015) indeed, nor are we in the grip of Eduardo Bonilla-Silva's (2014) 'color-blind racism'. Racism is not color blind nor is 'race' 'post'. In his paper, Ian Law addresses this issue by firmly locating the work of CERS within the long sociological tradition placing 'race' and racism at the center of the making of Western modernity, from Du Bois, Cooper, Césaire, and Fanon to contemporary theorists including Hall, Hesse, Collins, Goldberg, Glissant, and Winant. For him, it is important to keep the spotlight on racism as a primary field of research and practice in order to enable the global transformation of HEIs. At the curriculum level, this necessity is also highlighted by Ronelle Carolissen and Vivienne Bozalek's paper which draws on an interdisciplinary, inter-professional teaching, learning and research project set up across a historically disadvantaged (black) and a historically advantaged (white) HEI in Cape Town, South Africa, and across differently valued professions (Psychology, Social Work, and

Occupational Therapy) in order to address the historical and current racial inequities caused by apartheid's instantiation of racial difference and unquestioned white privilege irrespective of class. As these papers show, 'post-race' and 'color-blind' are pervasive institutional discourses which provide us with ways in which we can understand the insidious neoliberal racialization within which we find ourselves in the societies from which the papers in this issue draw. Pete Harris, Chris Haywood, and Mairtin Mac an Ghaill's article explores this neoliberal racialization by exploring the experiences of black and Muslim students by looking at how 'teaching otherwise' can create an alternative representational space. This space in turn enables a transformation in perspectives of self through pedagogy which is much needed in the future in UK HEIs if neoliberal racialization is to be effectively tackled.

Neoliberal racialization continues to be difficult to deal with because it is catalyzed by whiteness or 'whiteliness' (Yancy 2008, 2012) which are discursive and non-discursive aspects of institutional life which

> [...] becomes a deeply political, existentially *lived*, social category that shapes the subjectivities and future racialist/racist practices of whites. Whiteness is a way of performing both one's phenotypic white body/ one's subjectivity structured around a specific white racist epistemic orientation (Yancy 2008, 48).

The social body as skin, subjectivity, and epistemology are central to whiteness. As such, whiteness continues to be the motor of the egregious institutional racism which continues unabated even in the face of affirmative action programs. Joaze Bernardino-Costa and Ana Elisa De Carli Blackman look at the theme of the struggle against racism in Brazil and the adoption of affirmative action policies through the public universities of the nation because of the anti-racist actions of the 'movimento negro' (black rights movement). Affirmative action sprang from a Supreme Federal Court ruling in 2011 on the constitutionality of racially targeted policies in the University of Brasilia and the subsequent National Congress approval of quotas to be adopted by all federal universities in Brazil. However, even after much public debate, campaigning, and law making, the article shows that much still needs to be done, such as the adoption of affirmative action in postgraduate schools and in the contracting of teachers as well as the reconfiguration of the curriculum and of the research agendas of Brazilian universities. At this point in Brazil following the coup and the spread of conservative politics across the country seen in the recent local elections, there is increasing unease and much uncertainty about the future of quotas in debates from the Left. From the viewpoint of the USA, Gary A. Dymski looks at the institutional and specific disciplinary uptake of the diversity imperative and its successes and failures at the University of California Riverside (UCR) through its outreach, student support, and 'pipeline' programs. The strong performance of UCR in attracting and retaining students of color in 2014 led to its being ranked first in a poll of US universities meeting the 'Obama criteria' of access/ diversity/affordability/success. However, Dymski shows that much more still needs to be done into the future at both discipline and institutional levels as well as within political economy if students of color are to succeed in entering the professions.

The necessity for affirmative action policies illustrates that whiteliness is the bedrock of organizational culture and is embedded within institutional structures and processes as well as knowledge production and canonization which in combination enable racism 'to melt into thin air' (Gordon 1997). Whiteness works through a governmental (Foucault 1980) process of subjectification motivated by self-interest, personal benefit, and entitlement to undisputed privilege which Charles Mills (1997, 40) makes clear in the *Racial Contract*

> Both globally and within particular nations, then, white people, Europeans and their descendants, continue to benefit from the Racial Contract, which creates a world in their cultural image, political states differentially favouring their interests, an economy structured around the racial exploitation of others, and a moral psychology (not just in whites sometimes in nonwhites also) skewed consciously and unconsciously toward privileging them, taking the status quo of differential racial entitlement as normatively legitimate, and not to be investigated further.

Whiteliness is at the center of our putatively 'post-race' world and indeed has mythologized 'post-race' as a new form of 'racialized governmentality' which rules black, minority ethnic and white psyches, social spaces, and institutions alike. This is a racialized governmentality in which those racialized as non-white can be accused of racism against those racialized as white in a sleight of hand and perversion of knowledge and history which refuses white power and privilege as foundational to a description of racism. This is illustrated in Diane Watt's pedagogical focus on those 'difficult conversations' on racism aimed at enabling students to develop a critical understanding of the significance of anti-oppressive thought and practices. She found that when reflecting on anti-oppressive practices was made a core part of the curriculum this faced resistance from some white students who sought to undermine classroom debates about these issues effectively silencing those white students who wish to actively engage with anti-racist theory and practices. British black and South Asian students also felt marginalized by this resistance having to defend their experiences, or sometimes strategically avoiding the debates for fear of adversely affecting their relationships with some white students. Watt's paper powerfully illustrates the potentially contradictory outcomes of attempts at anti-racist practice within university teaching environments. Of course, this racialized governmentality is very little different from the evasive racism which Ruth Frankenberg's (1993) *White Women Race Matters: The Social construction of Whiteness* described in the twentieth century. 'Anyone can be racist' underlies racialized governmentality and must be critiqued as well as opposed as a mindset or perspective on the world if we are to change universities into workplaces which are not zones of toxic shock for faculty as well as into places of study in which students do not feel alienated.

What is interesting is that the pervasive power of whiteliness continues to be denied and indeed is balked at, remaining unsayable within universities. This regime of unsayability allied with the deniability of white power and privilege is why anti-racism has not worked. We cannot ameliorate something which we think does not exist because it is unsayable and deniable. Further, if we do notice and say 'this is racism' our acknowledgment is always tied to an individual failure or pathology on the part of both BME students and faculty and their white anti-racist allies. This culture of blame making means that we continually refuse institutional accountability for failure to address racism. Moreover, and much more insidiously, since the problem is constructed as that of those racialized as not-white and their allies racialized as white who continue to say that whiteliness is the root of the problem of continuing racial inequalities in universities, this claim falls on deaf ears. Such falling on deaf ears brings to mind Gayatri Spivak's (1995) subaltern who could never be brought into the scene of representation as recognizable political subject. Beyond the body racialized as black or minority ethnic, subalternity also continues to be the circumscribed space of anti-racist thought, practice and knowledge systems within UK universities.

Anti-racism has not worked as we can see in the continuing struggles for racial equality represented in the papers in this volume in societies in which 'race' continues to matter even though we might wish it were otherwise. Mark Christian's contribution highlights this persistence and its impacts at the level of the individual. His article speaks to black British male experience in US colleges and universities. It is an autoethnographic study in terms of relating, witnessing, and noting both learning and teaching experiences. The paper highlights the need for greater access and opportunity for black scholars to teach and study without stress and strain on their minds and bodies, especially for those facing the daily reality of teaching and researching within the context of Africana or black Studies in higher education. Christian notes that academia should be a place where liberal arts of all genres and their teachers are accepted and respected but there is still a long way to go before we can attest to the affirmative of this point of view.

Although saying anti-racism has failed fills us with feelings of political despondency, especially in the current UK context of BREXIT, failure must be acknowledged in order to build possible futures from the materials at hand in each country represented in this volume. The local is important to bear in mind because there cannot be a one size fits all approach to change even though we can say that we can learn from successes, failures, and hopeful shoots of change in each context. What we are talking about here we must remember is a very specific understanding of racism which has very specific Black Atlantic foci and approaches to its amelioration as we see from Mills (1997) above. What can we say though about anti-racism's failure within neoliberal institutions and neoliberal racialization?

Anti-racism's failure within universities

What institutionalized anti-racist policy and practice within institutions has done is to seek institutional transformation through changing structures and processes which militate against equality of access, process, and outcome because of the impact of whiteliness. This has basically been a liberal-inclusive approach based on a commitment to diversity which has not taken on board the pervasiveness of the Racial Contract. The Contract's pervasiveness is assured by the intensity of the affective attachment to privilege of those who benefit from it. It is further embedded within the psychic life of institutions and those who occupy and build them so that they can continue to occupy a world of instutitionalized racial inequality while chanting the 'post-race' mantra. In fact, to speak of being 'post-race' denies racism's contemporary existence (Goldberg 2015) and relegates it to a best forgotten past. It is interesting how one can say that racism does not matter while watching the events unfold which led to the 'Black Lives Matter' campaign in the US, or the shooting of Mark Duggan and its aftermath in the UK, or the continuing under-representation of Indigenous People in universities in Brazil and Canada, or the 'Rhodes Must Fall' campaign in South Africa. How can this will to silence continuing racism through asserting 'post-race' status in UK universities be explained?

By way of explication, let us turn again to the Racial Contract and the process of becoming white. This latter

> has nothing to do with a so-called genetic racial substratum, but everything to do with what happens at the level of social constitutionality, how the human being comes *to be* the white self that is both constituted by and constitutes white racism. (Yancy 2008, 48)

The process of becoming white is linked to the Contract which itself is based on keeping European and European descent white superiority in place for its signatories at the levels of political economy, culture, psyche, and epistemology. This ensures the continuation of racial exploitation and a normative position in which white privilege need not be questioned. Racism is silenced through what Mills (1997, 18) calls 'epistemologies of ignorance'. However, 'ignorance' does not mean 'unknowing' as we would expect from its etymology.

Rather, what we have instead are 'white misunderstanding, misrepresentation, evasion and self-deception on matters related to race' (Mills 1997, 19). In the twenty-first century what Mills' *white* 'mis' means is *not* that we live with white understandings, representations, evasions, and deceptions which are abnormal, bad, wrong, or divergent, all of which would be the normal understandings of 'mis'. Rather, what walks amongst us and stalks the halls of academic life is a *knowing ignorance* of whiteness and its racist impacts so that whiteness remains innocent of racism and unproblematically claims that space because of its 'unknowledges'. Whiteliness and white supremacy do not need to be defended against the charge of racism because of 'unknowledges'. 'Sometimes these "unknowledges" are consciously generated, while at other times they are unconsciously generated and supported (…) [but] they work to support white privilege and supremacy' (Sullivan and Tuana 2007, 2). 'Unknowledges' are linked to pervasive institutional racism through helping to maintain racism's deniability. These deniability regimes are crucial to the continuation of whiteness in universities through its racial affective economies and cultures of disattendability (Tate 2013), curricula, and interpersonal relationalities which lead to promotion or lack of it, student/staff experiences of racial privilege/disprivilege, and denial of access to the institution in the first place (Gutiérrez Rodríguez 2010, 2016). This is the weight of whiteliness which anti-racism has not managed to erase or even ameliorate even with all of the equality and diversity paperwork which exists in the different contexts examined in the papers here. This is its failure. It is not a failure produced by anti-racists but is one that was a direct result of its institutionalization and colonization as 'equality and diversity' after it had been stripped of its potential for critique and action. After all, it is impossible to allow unfettered institutional access to something which has such a fundamental critique of that from which you benefit and that which ultimately is not in your interest to change. We continue to struggle to name racism and to act against it within the university sector because of 'equality and diversity' as the preferred approach to racial inequity and institutional transformation.

Recognizing the basis of institutional inertia around racism or the erasure of past anti-racist changes leads us to now think about how to re-engage with the continuing necessity for anti-racist action in 'post-race' times. The question for the conference was 'Building the Anti-racist University: What next?' as it is for all of those who strive for racial intersectional equality. That 'next' is an important, indeed a vital shift, which will take us into thinking about how we can take forward the student campaign's call to decolonize the institution as our future option in the face of anti-racism's failure to make lasting and fundamental anti-racist changes to UK HEIs.

Decolonizing the university in 'post-race' times

What is it that we mean when we use this buzzword, what is it to decolonize this whiteness and white supremacy to which even those living with and through racial disprivilege can ascribe because of the pervasiveness of the Racial Contract? Let us begin from looking at what Glissant (1997) tells us about epistemological, societal, and self-liberation within his take on creolization as a rhizomatic movement which disrupts identitarian politics as it produces new subjectivities, a new 'common' (Hardt and Negri 2009) which recognizes white supremacy and racism as we break away from knowing unknowledges. Glissant locates the Caribbean archipelago as a zone of diversity which separates it from continental thought based on the One of universalism. His work makes us see the 'poetics of relation' within the decolonial moment as a break from the 'philosophies of the One of the West' (Glissant). The 'One of the West' here is whiteness whether read as psyche, institution, process, structure, affect, or political economy, for example. That is, Glissant enables us to continue to think about the project of decoloniality in terms of knowledge, power, becoming, and affect.

Let us begin to think the university as a contact zone, a zone of creolization which still continues to imagine itself as the place of imperial whiteness. Glissantian creolization is an ongoing relational process which inscribes the principle of non-hierarchical unity with a relation of equality with and respect for the other as different from oneself within a natural openness to other cultures. The principle of equality and respect for the other as different not inferior is crucial to the decolonial moment as it is through this that we can begin to prise open what Marley (1980) calls 'mental slavery', what Fanon (1986) would term the 'colonial psyche' and what Mills (1997) has called 'a moral psychology'. This lays out the necessity for psychic and epistemological decolonization which both looks at whites' and at racialized others' complicity in keeping the status quo in place because of the benefits that they feel they gain. Creolization, like decolonial thinking, does not universalize itself unlike the One of the West but 'brings into Relation' hitherto disparate constituencies (Glissant 1997, 90). Relation produces new identities through errantry, a psychic mode of affirming racial identities as an antidote for and in opposition to exile which can potentially erode one's identity (Glissant 1997, 20). Errantry builds a new racialized and racializing common as it includes both collective and individual in knowing that 'the Other is within us and affects how we evolve as well as the bulk of our conceptions and the development of our sensibility' (Glissant 1997, 27). This recognition of the fact of whiteness within us as individuals and communities is essential in decolonizing racialized psyches whether those are black, Indigenous, People of Color, or white as we build what Glissant describes above as a non-hierachical unity. A unity which for our purposes is an anti-racist common.

Decoloniality thus necessitates working at the levels of the manifestation of white racialized power and its attendant prescriptions of what counts as knowledge, who can occupy the category human and its negative affects which circulate and make HEIs the site of pain for both faculty and students alike. To include a consideration of affect as being in need of decolonization is an important addition to the trio of the coloniality of power, coloniality of knowledge, and coloniality of being which is the usual approach to decolonial thought (for example, Maldonado-Torres 2016). Resisting the coloniality of knowledge through demanding epistemological decolonization is an essential aspect of the decolonial project and it is not a happy coincidence that UK students have this firmly in their sights with the campaign 'why is my curriculum white?' This question has been a long time coming but is a significant one especially if one thinks about the 'post-race' context. That is, if 'race' does not matter only class, then why is there still a blinding whiteness in terms of what counts as knowledge, in terms of what has become the canon, what gets taken up, and what remains erased? What we now need is a necessary re-reading of 'post-race' which sees it as pointing only to the construction of a present and future time and space in which whiteness as 'race' power and privilege is erased, in which the anti-black/People of Color/Indigenous racism it generates ceases to exist.

For the first time in UK history and that of Europe, there is a black Studies degree in a university – Birmingham City University. This did not emerge at UCL-home of the Galton Collection and Galton Lecture Theater in memory of the man who first coined the term 'eugenics' in 1883 – even though its promise began there. This development is quite momentous and must be applauded as a response to the issue of the white curriculum. This does not take away from all the work which has been engaged in for years by colleagues at other UK institutions but begs the question of why the Russell Group as a whole did not make a similar response. Similarly, it is important to ponder why this innovation came from a new university in a multi-racial city like Birmingham with its rich black intellectual and activist history, including being the home of the now defunct Center for Contemporary Cultural Studies at the University of Birmingham. This makes us note the affects attached to white epistemology across the university sector where even now very few courses which look in a sustained and in-depth way at racism and black Studies exist and those that do are currently being dismantled. These whitely affective attachments create a connection between the white body irrespective of gender, class, sexuality, age, and location and the white epistemological tradition constructed as superior, whatever the discipline. Both bodies and epistemology attain value because of this connection so, of course, it is clear that a Black Studies program already sets into train a destabilization of these certainties. This inherent critique of the value of whiteness as body and knowledge is perhaps what led to the demise of many Black Studies programs in the 'post-race' US and what has led to the demise/diminution of those few courses that there were in the UK.

What has changed in the Higher Education sector to now enable the emergence of a Black Studies program at undergraduate level in the UK? Perhaps, it is that very same neoliberal racialization and commodification of knowledge to be sold to niche international and national markets which has enabled this development. Perhaps, everything is related to political economy in the end as the profit imperative in marketized UK universities necessitates the development of an international/national market in students willing to pay for a 'British education'. Ironically, marketization might be the motor which drives the development of curricula which attempt to be non-Eurocentric as it 'brings into relation' previously disconnected constituencies.

It continues to be necessary to draw together the issues emerging from the debates throughout the articles in this special issue on curriculum, pedagogy, access, policy, process, experience, outcomes, subjectivities, racialization, and racism in HEIs in Brazil, South Africa, Canada, the USA, and the UK to craft an agenda for building the anti-racist university into the 'post-race' twenty-first century in contexts where white privilege and power remain. These must be 'the next steps' but ones which are continuously reiterated and reinscribed as racism morphs because white privilege will continue to be maintained in the face of future decolonial assault.

Disclosure statement

No potential conflict of interest was reported by the authors.

References

Ahmed, Sara. 2012. *On Being Included: Racism and Diversity in Institutional Life*. Durham, NC: Duke University Press.

Bagguley, Paul, and Yasmin Hussain. 2007. *The Role of Higher Education in Providing Opportunities for South Asian Women*. Bristol: The Policy Press.

Bonilla-Silva, Eduardo. 2014. *Racism without Racists: Color-blind Racism and the Persistence of Racial Inequality in America*. 4th ed. Plymouth: Rowman and Littlefield.

Coard, Bernard. 1971. *How the West Indian Child is Made Educationally Sub-normal in the British Education System*. London: New Beacon for the Caribbean Education and Community Workers' Association.

Connor, Helen, Claire Tyler, Tariq Modood, and Jim Hillage. 2004. *Why the Difference? A Closer Look at Higher Education Minority Ethnic Students and Graduates*. Research Report No. 552. London: Institute for Employment Studies.

Equality Challenge Unit. 2015 Equality in Higher Education: Statistical Report.

Fanon, Frantz. 1986. *Black Skins White Masks*. London: Pluto Press.

Foucault, Michel. 1980. *Power Knowledge: Selected Interviews and Other Writings 1972–1977*. Edited by C. Gordon. Brighton: The Harvester Press.

Frankenberg, Ruth. 1993. *White Women, Race Matters: The Social Construction of Whiteness*. Minneapolis, MN: University of Minnesota Press.

Glissant, Édouard. 1997. *Poetics of Relation*. Translated by Betsy Wing. Ann Arbour, MI: University of Michigan Press.

Goldberg, David Theo. 2015. *Are We all Post-racial Yet?* Cambridge: Polity.

Gordon, Lewis Ricardo. 1997. *Her Majesty's Other Children: Sketches of Racism from a Neo-colonial Age*. Lanham, MD: Rowman and Littlefield.

Gutiérrez Rodríguez, Encarnación. 2010. "Decolonizing Post-colonial Rhetoric." In *Decolonizing European Sociology*, edited by Encarnación Gutiérrez Rodríguez, Manuela Boatca, and Sergio Costa, 49–69. Abingdon: Ashgate.

Gutiérrez Rodríguez, Encarnación. 2016. "Sensing Dispossession: Women and Gender Studies between Institutional Racism and Migration Control Policies in the Neo-liberal University." *Women's Studies International Forum* 54: 167–177.

Hardt, Michael, and Antonio Negri. 2009. *Commonwealth*. Cambridge, MA: Harvard University Press.

Maldonado-Torres, Nelson. 2016. "Fanon e Foucault sobre Modernidade, Poder, Conhecimento, e a Zona do Não Ser [Fanon and Foucault on Modernity, Power, Knowledge and Being]." Paper presented at International Seminar on a Black Perspective on Decoloniality, University of Brasilia, Brazil, October 5–7.

Marley, Bob. 1980. "Redemption Song". Uprising album.

McManus, I. C., A. Esmail and M. Demetriou. 1998. "Factors Affecting Likelihood of Applicants Being Offered a Place in Medical Schools in the United Kingdom in 1996 and 1997: A Retrospective Study." *British Medical Journal* 317: 1111–1117.

Mills, Charles. 1997. *The Racial Contract*. Ithaca: Cornell University Press.

Spivak, Gayatri. 1995. "Can the Subaltern Speak?" In *Colonial Discourse and Post-colonial Theory: A Reader*, edited by P. Williams and L. Chapman, 66–111. Hemel Hempstead: Simon and Schuster International Group.

Stone, Maureen. 1981. *The Education of the Black Child in Britain: The Myth of Multi-racial Education*. London: Fontana.

Sullivan, Shannon, and Nancy Tuana. 2007. "Introduction." In *Race and Epistemologies of Ignorance*, edited by Shannon Sullivan and N. Tuana, 1–12. Albany, NY: State University of New York Press.

Tate, Shirley Anne. 2013. "Racial Affective Economies, Disalienation and 'Race Made Ordinary." *Ethnic and Racial Studies* 37 (13): 2475–2490.

Yancy, George. 2008. *Black Bodies, White Gazes: The Continuing Significance of Race*. Lanham, MD: Rowman and Littlefield.

Yancy, George. 2012. *Look a White! Philosophical Essays on Whiteness*. Philadelphia, PA: Temple University Press.

Race, racialization and Indigeneity in Canadian universities

Frances Henry, Enakshi Dua, Audrey Kobayashi, Carl James, Peter Li,
Howard Ramos and Malinda S. Smith

ABSTRACT

This article is based on data from a four-year national study of
racialization and Indigeneity at Canadian universities. Its main
conclusion is that whether one examines representation in terms
of numbers of racialized and Indigenous faculty members and their
positioning within the system, their earned income as compared
to white faculty, their daily life experiences within the university as
workplace, or interactions with colleagues and students, the results
are more or less the same. Racialized and Indigenous faculty and the
disciplines or areas of their expertise are, on the whole, low in numbers
and even lower in terms of power, prestige, and influence within the
University.

Introduction

Over the past several decades, Canada has become increasingly ethnically and racially
diverse and the Canadian Indigenous population has grown significantly, yet racialized
and Indigenous peoples are underrepresented in major institutions. A significant body of
research and scholarship on equity and diversity in higher education has documented the
persistence of systemic barriers and implicit biases faced by members of equity seeking
groups – women, racialized minorities, Aboriginal[1] peoples, and persons with disabilities
(Carty 1991; Mukherjee 1994; Monture-Angus 1995, 1998; Razack 1998; Dua and Lawrence
2000; Prentice 2000; Dua 2009; Henry and Tator 2009; Smith 2010). Despite the expanding
research on equity and higher education, analyses of racism, racialization, and Indigeneity in
the academy are notable by their absence. No major scholarly body – whether representing
universities, presidents, deans, or university teachers – has given priority to the implications
of the cultural heterogeneity in higher education, and none has undertaken a study of the
status and everyday lived experiences of racialized scholars and scholarship in the academy.
Despite many efforts, which most often amount to no more than well-worded mission
statements and cosmetic changes, inequality, indifference, and reliance on outmoded conservative traditions characterize the modern neoliberal university.

Using data from our recent nationwide study *Race, Racialization and the University* which foregrounds *racism* as a critical variable shaping peoples' lives and experiences, we examine what universities have done, and question the effectiveness of their equity programs. We also set out the experiences of racialized faculty members across Canada for whom strong claims of equal opportunity have not really changed their everyday working conditions.

Methodology

We employed a multileveled and mixed methods approach – census data (statistical analysis using Public Use Micro Files as well as Research Data Center original data), surveys, interviews, textural, and policy analyses. Our methodology utilized the strength of qualitative and quantitative approaches. To gain an overall picture of the university faculty population as well as their earnings, a questionnaire survey administered in eight universities,[2] and interviews with 89 racialized and Indigenous faculty, equity directors, and administrators were conducted in 12 universities selected on the basis of size, region, and interest in the subject matter. Interviewees were secured through personal contacts and snowballing techniques increased our sample size. Interviews were guided by pre-constructed questions and conducted informally ensuring confidentiality. Faculty members were generally eager to speak of their experiences; for many, this was cathartic since they rarely discussed racism.

Going beyond a focus on numerical representation meant looking at everyday experiences with racism, the ways in which institutions create an understanding of equity, and the effectiveness of the mechanisms to address inequities. Therefore, we examine the multiple and interrelated ways in which racialization and racism take place by analyzing data on: (1) representation relating to hiring, tenure and promotion practices, and the attitudes and practices of administrators responsible for equity policy and practice; (2) institutional/ organizational culture that generates barriers to access and equity; (3) mechanisms for inclusion, noting what universities have put in place to ensure inclusion; and (4) discourses in terms of the social construction of knowledge about equity, diversity, inclusion, and exclusion and how these have been used by the academy to inform its practices.

This study is the first national study to address the status of racialized and Indigenous scholars in Canadian universities. Until this point, the Canadian literature had focussed on either case studies of one university or experiential analyses written by Indigenous and racialized faculty. As important as these studies have been in highlighting patterns of racism, a national picture was missing and, in fact, no such study seemed to exist in the international context. We supplemented our national analysis with more detailed study of a sample of 12 Canadian universities which represents a diversity of regions and institutions. As a result of its scope, this study has gathered extensive data in order to make as accurate as possible an assessment of the position of racialized minorities within Canadian universities.

We encountered some difficulties in measuring representation of racialized and Indigenous faculty, mainly due to lack of disaggregated data. Inter- and intra-group differences with respect to gender or other markers of difference were impossible to assess (Jayakumar et al. 2009). The lack of data affected both the quality of research findings and the conclusions that can be made. Difficult as it was to obtain good data on those employed within academic institutions, the data presented here are as significant for those whom they do not describe as they are for those described. The issue is not simply one of obtaining more data, but of asking who is included and why (Dua and Bhanji 2012). Notwithstanding differences in the ways in which the academic work force is categorized, it cannot be denied that under-representation occurs, that women are less represented than men, and that there are significant differences in the numbers and the patterns of representation of different racialized groups. Under-representation points to obdurate barriers to access and participation of racialized and Indigenous academics.

The context

From the perspective of racialized and Indigenous faculty members, we examine whether institutions seem ready to accommodate not only their presence but also their scholarship, pedagogy, service inclinations, and cultural and social capital shaped by their communities. We ask, what life is like for racialized and Indigenous faculty members in universities shaped by neoliberal individualism, merit, competition, and entrepreneurship (Kurasawa 2002; Luther, Whitmore, and Moreau 2003; Mahtani 2004; Newson 2012; Thornton 2012; Griffin, Bennett, and Harris 2013; Giroux 2014; James and Valluvan 2014).

Drawing on qualitative data of the experiences and perceptions of racialized and Indigenous faculty, we use the prisms of critical race theory (CRT) and whiteness, employment equity, and neoliberalism to examine how the social, political, and cultural climates of their institutions have enabled or limited their role as agents of change, and what their presence has meant in helping to advance equity in their universities. Scholars suggest that the seeming shift over the last four decades toward more accessible and inclusive universities corresponds to the neoliberal shift in society as a whole – which has operated not only to demoralize faculty members, but also to obfuscate the university's shared responsibility (Kurasawa 2002; Luther, Whitmore, and Moreau 2003; Ahmed 2012; Newson 2012; Thornton 2012; Giroux 2014). In a context in which the ideologies of neoliberalism and whiteness structure the articulation and evaluation of merit, democracy, and diversity (in both membership and scholarship), racialized and Indigenous faculty members tend to experience work situations where they have limited control over their working conditions, institutional barriers to their scholarly potential and productivity, and challenges to their professional judgements and entitlements – factors that are typically associated with a precarious work situation (see Braedley and Luxton 2010; Thomas 2010; Law Commission of Ontario 2012).

Disciplines and the departments or programs that host them often function as gateways to the academy. They may open doors but they may also put up walls and police boundaries in ways that limit access and change and, thereby, conserve the prevailing order. In order to advance equity, diversity, and complexity in the university, more attention needs to be focused on disciplines as a unit of analysis and the ways they reflect and represent historical and social realities such as diversity and decolonization. Canadian society is undergoing a fundamental demographic transformation. Despite decades of talking about equity, diversity, and inclusion in society and the academy, this demographic transformation is not reflected in the academy and the absence is especially notable in the composition of faculty and leadership, which remain overwhelming white and primarily male. The invisibility of broader representation of diversity also remains evident despite almost three decades of self-studies, which until recently have narrowly focused on the status of women. Where disciplinary diversity is evident, in hiring or teaching and research, it is primarily in the area of women, gender, and sexuality studies. This means Indigenous and racially and ethnically diverse students in many social science and humanities disciplines, in particular, never or rarely experience someone like themselves as university professors, mentors, and leaders, and as researchers and knowledge producers.

In proceeding, we discuss how the tenets of neoliberalism and whiteness structure how universities respond to perceived needs for equity programs. We first examine the policies that frame 'equity' and 'representation', noting the results of those programs in terms of measurable aspects, that is: increases/decreases in representation, and variation in salaries. We then address the precariousness of racialized and Indigenous faculty members' work situation using their own assessments from surveys and in-depth interviews. We discuss their perceptions of and experiences in terms of how they are positioned in the university, and the extent to which the climate in which they work opens up or limits scholarly research, teaching, and service opportunities. Finally, we address the process of racialization itself, examining the ways in which everyday events in the university create racial difference and oppression. Three main concepts underlie our research: CRT and whiteness, employment equity, and neoliberalism.

CRT and whiteness

The project is informed by CRT (Crenshaw et al. 1995; Delgado and Stefancic 2012), including whiteness studies and intersectional thinking. Bell (1980) stressed that the systemic oppression of African Americans, and by extension Black and racialized people in many areas of the world, cannot be understood without reference to how capitalism, the free market economy, the political status quo, and other conservative institutions maintain white privilege. Institutions of white privilege must be acknowledged if the rights and interests of non-whites are to be fully recognized.

CRT scholars deconstruct the assumptions that, when posited as 'universal', form the foundation for white privilege and power. CRT challenges antidiscrimination policies that do not take into account the linkages between race, class, and gender, which structure the everyday racialized experiences of Indigenous and racialized people as they engage with sectors and systems such as education and the media (Williams 1992; Ladson-Billings 1998; Dua and Lawrence 2000; Monture 2010). It also emphasizes the role that narrative and storytelling play in analyzing the nature, dynamics, and impact of racism. Victims' stories help us to understand feelings, perceptions and experiences, interpret myths and misconceptions, deconstruct beliefs and common-sense understandings of race, and unpack the ahistorical and often decontextualized nature of law and other 'science' that renders mute the voices of the marginalized group members. The role of 'voice' is central to a critical race approach (Henry and Tator 2009; Smith 2010).

Whiteness Studies is closely aligned to CRT. It focuses on how white skin confers privilege systemically and structurally while excluding racialized people from the benefits of society. The category 'white' is socially constructed, and operates in relation to 'whiteness', which 'refers to a set of assumptions, beliefs, and practices that place the interests and perspectives of white people at the center of what is considered normal and everyday' (Gillborn 2015, 278). Both whiteness and blackness are racialized. Whiteness studies racialize the white race and uncovers the ways in which white privilege is unconsciously acquired and exercised. White privilege is transnational and comes from the history of European imperial and colonial expansion and its continuing legacies globally. Some of the commonly held discourses labeled 'discourses of domination' by Henry and Tator (2009) include the myth of color-blindness, in which people are assumed not to recognize skin color as a racial differentiating trait in making decisions. Gotanda (1991; cited in Vaught 2011) criticizes the assumption that people do not 'recognize' constructs of race in making decisions and argues that such non-recognition 'fosters the systematic denial of racial subordination and the psychological repression of an individual's recognition of that subordination, thereby allowing such subordination to continue.' In other words, non-recognition of race permits the continued opacity of white privilege and domination.

Another analytical concept that has relevance to the present study is 'intersectionality,' as there are many forms of inequality that interact with one another, and individuals and groups have multiple, interacting identities. Race intersects with gender, class, disability, and other social and demographic characteristics to shape social and economic experiences. The concept, originally proposed by legal scholar Crenshaw (2002), is one that 'goes beyond conventional analysis in order to focus our attention on injuries we might otherwise not recognize … to (1) analyze social problems more fully; (2) shape more effective interventions; and (3) promote more inclusive coalitional advocacy.' In our research, we gave attention to gender and its intersectional relationship with differences in income, ethnicity, and daily lived experiences in the lives of racialized faculty. With respect to social class, Solomos has recently reiterated that class hierarchy is still fairly evident in the United Kingdom. The persistence of inequalities is primarily a function of the failure of the state to ease the erosion of the working class (BSC Conference 2015). Class and increasingly immigration have become substitutes for what he calls 'color coded' racism. These factors underlie the role of racism and become the major focus of government intervention.[3]

Along with other critical race scholars, we see that intersectionality has both an empirical and an activist component. It is a tool for analyzing related forms of oppression which aims to resist and challenge the status quo's denial of equality. Yet it has become a mantra in some social science literature to the extent that single variable analysis is criticized for ignoring or paying less attention to multiple forms of oppression (Gillborn 2015). One of the founders of CRT, Richard Delgado, recently noted that intersectionality can be taken to such extremes that it becomes paralyzing, 'because of the realization that whatever unit you choose to work with, someone may come along and point out that you forgot something' (cited in Gillborn 2015, 279)

Gillborn (2015, 277) notes that

> any attempt to place race and racism on the agenda, let alone at the *center* of debate, is deeply unpopular. In the academy, we are often told that we are being too crude and simplistic, that things are more complicated than that, that we're being essentialist and missing the *real* problem – of social class.

While it is fruitless to contest the role of social class in any analysis, we need to guard against subsuming race and racism within a class analysis since the attitudes, perceptions, and stereotypes that underpin racism can be found at any class level. Indeed, we recognize that subtle and elusive forms of 'othering,' leading to discrimination and marginalization, are the twenty-first Century's primary form of racism in many institutions and societies. This racism pervades all social institutions and social classes therefore focusing on racism in universities makes good empirical sense. Despite increasing diversity, including students who come from the poorer and working class sections of society, universities are still seen as a middle-class institutions which provide pathways for social mobility. While overt forms of racism are largely, although not exclusively, attributed to lower social classes and associated with economic competition, middle-, and upper-class racism is far more sophisticated and complicated. While many people are appalled to hear that racism exists at universities – the highest seat of learning – it is because they conceive of racism in its overt forms. They cannot comprehend that, for example, that criteria for the denial of tenure based on publishing in the 'wrong' journals or not bringing in enough grant money are manifestations of racism when its objects are people deemed to be 'different.' Intersectionality is vital to our framework and as critical scholars of race and racism we also believe that there is a need to focus on racism and its many elusive forms.

Employment equity

It is now more than three decades since the 'Abella Commission' (Royal Commission on Equality in Employment 1984) introduced the concept of Employment Equity, a made-in-Canada term intended to address barriers to entering the workplace and conditions in the workplace. Abella named four groups – women, persons with disabilities, Indigenous peoples, and members of visible (or racialized) minorities – that would be designated under Employment Equity legislation. Her recommendations were largely responsible for the 1986 adoption of an Employment Equity Act that initially applied to all federally regulated employers of a certain size. A decade later, the legislation was amended to incorporate the federal government itself and federally regulated employers under one Act, and to strengthen the planning and reporting dimensions of the policy. Within that rubric, universities fall under the Federal Contractors Program, which ties eligibility for contracts to a requirement to file reports and to set targets on equity hiring. Employment Equity programs were established in most Canadian universities in the 1990s, aimed at removing structural barriers and biases that hindered the recruitment, hiring, tenure, and promotion of racialized faculty (Dua 2009).

The program has had limited regulatory function and, over the past decade of Conservative government, Industry Canada has stopped monitoring altogether and does not provide any data derived from the annual reports. The university – or any workplace for that matter – does not exist in isolation, and Employment Equity is therefore not something that the university can achieve alone. Our focus on employment *equity*, therefore, emphasizes the need for specific measures to ensure equitable outcomes.

Neoliberalism

Our third major concept, neoliberalism, refers to the instrumental governmentality of the late 20th and early 21st Centuries. The term is rooted in monetarist policy of the Thatcher-Reagan era, with their emphasis on shrinking the state, the freedom of the market, the privileging of (certain) entrepreneurial ideas, and the rollback of government spending on social programs and policies. Neoliberal approaches include the withdrawal of governmental support for programs such as accessible education or social housing, as well as the underfunding and de-funding of grassroots and nongovernmental organizations. At the same time, we have stronger governmental intervention through an emphasis on audit culture, narrow notions of accountability and fiscal oversight, and stronger support for private business.

In Canadian universities, the rise of neoliberalism and audit culture have led to the adoption of managerialism, and widespread use of performance indicators and benchmarking. It has meant shifting funding to research projects that have the support of private businesses, and shifting university resources from supporting faculty to creating larger accounting departments with ever stronger audit requirements. All of these developments profoundly affect who gets hired and what they do, how many students they teach, how their time is apportioned, and what kinds of support and respect they receive for their research. More and more of the curriculum is taught by contract faculty rather than those in tenure-track positions. There is substantial evidence that members of the equity-seeking groups are disproportionately affected by neoliberalism.

Our findings

Representation, income differentials: the survey results

Our analysis of census data reveals that racialized and Indigenous professors are not only under-represented in universities (a situation which worsened over time);[4] they also earn lower wages than do their white counterparts, even after controlling for variables such as years of service and academic level. These earning differentials points to a number of questions: If racialized and Indigenous professors have the qualification and human capital needed to become professors, why are they not hired at the same rate as white professors? Some people have argued that they might earn a degree but underperform in other criteria evaluated in hiring, such as publication record, lack of success in funding, and providing appropriate service to the university and community. But the onus is on those who use this argument to present convincing data. Differences in income can also be justified according to similar criteria. If racialized and Indigenous professors are less productive than their white counterparts, then this is used as justification for lower remuneration. For such an argument to hold, however, convincing data that indicate their systematic underperformance in productivity compared to white professors at all age levels are needed. At the very least, our data clearly suggest that racial inequality in representation and earnings cannot be easily dismissed by productivity differences alone.

In our national questionnaire survey of eight universities in English Canada, we found a higher number of men than women among racialized and Indigenous faculty, with the vast majority of racialized faculty (two-thirds) identifying as immigrants.[5] They disproportionately worked in Medicine/Dentistry, Engineering, and Science/Computer Science and not the Arts and Humanities; and they have worked fewer years in the academy. This pattern points to evidence of a racialized-segmented-academic-labor market in Canadian universities. While there were only eight universities enumerated in our sample, it is likely that the trend extends more widely. It is clear that far more needs to be done to diversify the entire university and not just a small number of faculties. Canadian universities and their students need more racialized professors who teach in the Social Sciences and Humanities in addition to those already teaching in Engineering, Medicine/Dentistry, and Science/Computer Science. Their perspectives can help change the social and cultural narrative of Canada to one that better reflects an increasingly multi-racial/cultural/ethnic population.

With regard to the productivity of racialized faculty across various disciplines, we found that they are 'playing the game.' Racialized faculty outperform their non-racialized counterparts in winning research grants and publishing articles but have few book chapters and books. It is worth noting, as heard through our interviews, that these faculty members kept up this publication record even as many were told that their research was too political, too ideological, or too rhetorical. Further, an examination of these faculty members' tenure and promotion found that racialized faculty were less likely to be awarded these benchmarks, but if they do manage to earn them, there is marginal difference in how long it took them. The survey also showed that racialized faculty members perceive tenure and promotion to be influenced as much by 'soft' metrics such as personality, civility and collegiality, as by 'hard' metrics like publication and winning grants. The opposite pattern is largely found with perceptions about administrative and committee appointments and hiring. Few racialized faculty agreed that equity considerations were factors affecting hiring, tenure, and promotion, as well as appointment to administrative and committee appointments. As other aspects of our study found, these findings suggest that equity policies are not working and racialized faculty are aware of this failure. We further analyzed the perceptions of work load, to find marginal differences in perceptions between racialized and non-racialized faculty.

When asked about perceptions of tenure and promotion, 'hard' metrics of performance appear to be undervalued by racialized faculty. This could be because of the tensions between high rates of output and lower rates of reward for them. This is also illustrated by the higher rates of agreement by racialized faculty on the importance of 'soft' metrics of performance, those that are least quantifiable and observable empirically. It appears that racialized faculty recognize that their academic output or production might matter less than affinity and network biases, such as who they know and how they get along with them. This pattern might reflect a pragmatic outlook on the devaluing of their labor and skills. Differences between racialized and non-racialized faculty members' perceptions of the role of 'hard' and 'soft' metrics are also seen in administrative and committee appointments. For the former, racialized faculty appear to prize 'hard' metrics more than do non-racialized faculty, which may mean they have confidence in the academy once they have broken barriers into it. This suggestion is in line with the findings on the differential pathways of those who achieved tenure and promotion. In contrast, racialized faculty were far more ambivalent and skeptical of factors that affect hiring, perhaps reflecting a malaise associated with the leaky pipelines and blockage we found in other data reported above. The skepticism racialized faculty express with regard to hiring might be best illustrated with the low level of agreement that equity considerations play a role in hiring, despite Employment Equity policies that shape all Canadian university job ads.

As a whole we find that racialized faculty understand the Canadian academic system and 'play the game.' That is, they have the human capital and demonstrate a high level of performance on outcomes that should be rewarded by universities; however, their perceptions of how to best navigate that system are clearly different from those of their non-racialized colleagues. Such differences in perception are very much in line with previous research on perception of discrimination in the Canadian academy (Nakhaie 2004, 2007; Henry and Tator 2012). We believe that differences found among racialized faculty generally reflect a pragmatic and skeptical outlook on the Canadian academic system, which shows that some racialized faculty successfully navigate the system, but perhaps through a solitude of experiences that their colleagues fail to see.

University reponses to inequities: anti-racist initiatives

Our study illustrated that equity initiatives are unevenly developed. Universities vary substantially in the kinds of policies available to address inequities and racism, the mandates of the office responsible for dealing with discrimination, reporting structure, and number of staff.

We found that three dominant frameworks are deployed to address inequity – human rights, equity, and diversity frameworks. These frameworks differ in how they address racism. Human rights frameworks focus on implementing government requirements that employers have anti-harassment and anti – discrimination policies. Equity frameworks employ a broader mandate to address systemic discrimination, while diversity frameworks, emerging as a backlash against equity frameworks, are seen as less conflict ridden. Most universities have some infrastructure in place to implement policies. Thirty-five of forty-nine universities had developed dedicated offices that are directed to address harassment in the workplaces, and enhance equity. These offices often focus on faculty and staff concerns, leaving student issues to be dealt with by Ombudspersons, Student Services, or Deans. These offices vary substantially in the number of staff, and in the reporting structures. Finally, we found a proliferation of equity services, particularly in the larger universities, where senior administration appointments mandated to address equity have emerged, in addition to faculty equity offices.

In dealing with the effectiveness of equity policies in Canadian Universities, we found a broad range of mechanisms that addressed harassment, discrimination, and inequities to some extent, but all were assessed as ineffective in addressing racism. Formal processes for anti-harassment and anti-discrimination polices were riddled with ineffective procedures and resulted in conflict-ridden and unsatisfactory results. In addition, this mechanism was least able to address situations of racial harassment, bullying and discrimination. As a result, most universities attempted to resolve incidents informally. Informal mechanisms were also reported to be ineffective, however, in dealing with racism, racial harassment, and bullying. Educational workshops, despite their inability to reach most members of the university community, were assessed to be effective in shifting some aspects of institutional culture – but changing the influence of 'whiteness' was still seen as a challenge. Equity committees were effective in raising concerns about inequities and proposing remedies, but as these committees did not have mandates to ensure implementation, often their efforts were for naught. Equity plans were often put forward with little consultation, and not always enforced. As a result of the ineffectiveness of these mechanisms, in many universities, senior administrators are being mandated to oversee equity. This strategy was assessed to be the most effective in furthering equity, as well as ensuring a systemic approach in which different constituencies are accountable for equity; however, senior administrators reported that resistance to their efforts limited their success.

The ineffectiveness of human rights and equity mechanisms to address racism raises serious questions. Given the expansion of efforts to address equity, why are such efforts not more effective? This question is particularly pertinent as changes that could allow these mechanisms to be more effective were identified including procedural rules and mandates, more resources, greater input from equity activists, greater monitoring, and more administrative support. Why are these mechanisms that seem to be effective in addressing other inequities so ineffective in addressing racism?

The expansion of equity policies may have occurred in order to address other forms of inequities, such as sexism, gender, sexual orientation, and disability, rather than Indigeneity and racism. The 'culture of whiteness' makes it difficult not only to remedy incidents of racism, but also to shift the culture of academia so that such incidents would not occur. We found that the effectiveness of each mechanism was limited by the ways in which whiteness is structured and at the same time invisibilized in university settings. Perhaps the most significant aspect of whiteness is the power of white subjects to resist anti-racist efforts. Such mechanisms, rather than facilitating, inhibit the recognition and remedy of racism. Thus, our findings strongly suggest that the attention paid to equity is not necessarily tied to a commitment to addressing racism. Senior administrators in particular pointed out the results of such mechanisms are more 'performative' than substantive, thus obscuring the ongoing racism within higher education. Our findings resonate with Ahmed's (2012) research in England which draws upon Butler's concept of performativity as 'the reiterative and citational practice by which discourse produces the effects that it names.' She describes British academic context characterized by audits, embedded in diversity policies that provide apparent evidence of proactive measures, but concludes that diversity is '[i]n the world of the non-performative, to name is not to bring into effect' (Ahmed 2012, 117).

The complex set of interrelated factors that account for the expansion of equity – legal obligations, fear of litigation, fear of negative media coverage, the increasing competitiveness for students, especially international students, the importance of international reputations – also ties equity mechanisms to non-performativity. Moreover, the ineffectiveness of such mechanisms cannot be divorced from the complex relationship between managerialism, neoliberalism, and equity, where neoliberal tenets of competitiveness, markets, and efficiency have encouraged equity initiatives with limited scope. Thus, under neoliberalism, the most important measure of an equity policy is not its ability to address racism (and other forms of inequities), but rather to what extent the presence of these mechanisms leads to the perception that such universities are efficient, competitive leaders. Thus, the ineffectiveness of equity policies is not a failure but a very successful discursive act. Not only do these policies serve to mask discrimination, they offer a discursive non-performative process of naming 'not to effect.'

What our respondents said: interview findings

The overall picture is of a significant group of faculty across many Canadian universities with deep-seated and profound criticisms of the academy, its structures, and its governance. The university is largely perceived to be a traditional white and male-dominated institution that is taking only minimal steps to provide an inclusive welcoming environment for its racialized and Indigenous faculty.

In reflecting on current hiring practices one faculty member commented that someone like him 'would never be hired these days.' For the most part, this comment sums up the sentiments of many racialized and Indigenous faculty members we interviewed about their experiences in the academy. Given their experiences, they reckoned that the 'good intentions' of universities that brought them onto campuses in the 1980s were mere 'rhetoric' without substance – without the necessary institutional policies and practices that created an environment where the different knowledge and perspectives of Indigenous and racialized scholars were accommodated and incorporated. On this basis, they reasoned that universities were 'starting to become less progressive,' in that they were reluctant to have more racialized and Indigenous scholars on their campuses. As Giroux (2014, 17) argues, today's universities operate within a culture in which the academics best able to survive its challenges are those most comfortable with 'the corporatization of the university and the new regimes of neoliberal governance' where they are 'beholden to corporate interests, career building and the insular discourses that accompany specialized scholarship.' Many racialized respondents argued that only specific types of knowledge are recognized as legitimate, and Indigenous faculty members talked of their decolonial struggles to re-center Aboriginal history, philosophy, and culture and to incorporate anti-racist models of knowledge. In doing so, they are often met with deep resistance from white students, colleagues, and administration. Racialized faculty commonly experience demands from minority students wishing to have mentors and role models whom they believe can relate to them and their lived experiences.

The Indigenous and racialized faculty members also told us about sitting in hiring committee meetings and being part of conversations where they observed affinity, network and accent biases such that who one knew (or her/his network of friends) and accent (having a 'foreign' accent was considered a problem because students will not be able to understand the speaker) operated as invisible barriers to faculty appointments. And there were also the questions about these scholars' 'foreign' credentials if not obtained in Europe or North America, their ability to secure research and program funding, and their capacity to live up to expectations and images as representatives of their ethno-racial groups. These demands, questions, and expectations, combined with the absence of mentorship, the problematic relationships with their white colleagues and students, the insecurity generated by the tenure and promotion processes, and their struggles to be taken seriously and gain respect, contribute to precarious working situations and social relationships in which Indigenous and racialized faculty members found themselves, as well as the psychological state of ambivalence, skepticism, uncertainty, low self-esteem, hopelessness, and anguish they felt in their job.[6]

The presence of some Indigenous and racialized faculty disguises the fact that there has been little or no change in the ways institutions operate. Some faculty members increasingly feel marginalized in university environments that appear to be reverting to a traditional, white, homogeneous character, even as the university advertisements declare commitment to having an ethnically and racially diverse faculty body and affirmative action appointment committees are charged with implementing equity policies. Racialized and Indigenous scholars are often called upon to mentor a diverse student population, but such work taxes their time, and it is clear that their numbers are insufficient to address the needs of a future generation. Further, conscious of the fact that universities insist on having 'academic stars,' many racialized and Indigenous scholars are working diligently to prove themselves worthy of their tenured appointments and, in some cases, to prevent the demands of teaching and service (including that to their own communities) from interfering with their scholarship and limiting their productivity. But the reality is: the sometimes new, emerging, and different scholarship in which many racialized and Indigenous scholars engage has yet to earn the recognition and credibility. Conceding that they tend to be viewed as 'representatives' of their communities and as narrowly concerned with issues of equity, these faculty members acknowledged that how their behaviors and scholarly output are read have implications for the future of equity policies and practices, particularly with regard to hiring faculty members from their communities. Theirs is a precarious work situation where they constantly struggle against marginalization, racialization, tokenization, ghettoization, and alienation expressed in the demands that they conform, fit in, be star scholars, and meet an 'academic standard' that devalues the critical and transformative knowledge they bring to scholarship and the institution.

Conclusion

> If universities can't figure out how to deal constructively with our differences then you just have to give up hope generally. If we can't do it in universities then what hope does the rest of our society have? (interviewee)

Four decades of equity policies have failed to transform the academy significantly to make it more diverse and reflective of the broader society and student body. In part, this is because of structural barriers and discriminatory practices that have functioned to exclude and stall transformation. It is also a result of the inadequately examined preference for sameness that leads to practices of replication. Change has also eluded universities because of the subtle workings of unacknowledged biases that privilege affinity and the needs of dominant insider groups. Unconscious biases have a significant impact on the career trajectories of racialized and Indigenous scholars and women in the contemporary academy. The cumulative biases and structural barriers mapped along a spectrum or pipeline make visible the challenge for racialized and Indigenous faculty not only at the point of entry but, potentially, at every major stage of their academic careers. The biases tell a story about a potential obstacle to career mobility that many racialized and Indigenous scholars face. The complex dynamics of subtle biases and structural barriers also make visible how much harder they have to work in order to thrive and succeed in the academy. The findings suggest that biases and assumptions of whiteness have exacted an incalculable cost for many racialized and Indigenous scholars. They rob the academy and the broader society of a wealth of talent and the invaluable heterogeneity of people, their knowledge, and the perspectives that could make universities more equitable, diverse and excellent.

It is important that faculty members, and those hired with diversity in mind, enter an institutional climate in which their presence is valued for the new, additional, and different experiences and perspectives that they bring through their research questions and analyses, contributions to curricula, pedagogy, service, and scholarly activities (including their work with communities). If post-secondary institutions are to be relevant as knowledge generating entities in today's diverse communities, then they must be ready to accept and accommodate the presence of racialized and Indigenous faculty. Indeed, as Professor Joanne St. Lewis contends regarding the failure of universities to 'face up' to issues of race, 'it's not just how many black faces [there are] in the room. It's about the space to engage in intellectual work. It's about the opportunity to create new forms of knowledge' (cited in Drolet 2009).

Despite talk about an inclusive curriculum, and demands from Indigenous movements aimed at indigenizing, decolonizing, and internationalizing the curriculum in the westernized university, the scholarship on race/ethnicity, Indigeneity and, to a lesser extent, gender, remain on the margins of teaching and learning. Many students can graduate from a degree program and never grapple with issues of racism and decoloniality. This issue is particularly true of graduate programs engaged in training the next generation of scholars It also means that graduate programs are not providing new scholars the tools they need to grapple with colonial history, its relationship to power and the hegemony of imperial and colonial narratives that have established the terms of, and tools for, conversation in the westernized university.

If universities are places into which racialized and Indigenous people will gain access, fully participate, attain tenure and promotion, and have their scholarship recognized, then critical attention must be given to how the existing diversity discourse sustains color-coded power relations, inculcates expectations, and conveys reminders (e.g. 'you know why you were hired' – to represent 'your' communities and take care of marginalized students' and communities' concerns) that racialized and Indigenous faculty are not recognized or accepted as legitimate members of the academy with all the earned rights and privileges. What is clear, despite their increasing presence on university campuses, is that racialized and Indigenous faculty continue to struggle against their historical exclusion, and to justify the special measures, however limited, that have been implemented to make their presence in the academy possible. Frequent reminders of the reasons for their presence in the academy mean that they will constantly have to struggle against not only their own erasure, but also that of the faculty members they mentor, the students they teach, the research they conduct, and the scholarship they produce. That silence about race and racial issues remains the norm and does nothing to address the reality that race and racism have shaped and continue to shape the experiences, opportunities, and perceptions of racialized and Indigenous scholars. If the challenges that these faculty members face are to be effectively addressed, then, an institutional commitment to equity is integral to creating a welcoming and supportive academic culture.

Notes

1. Federal Government documents use the term 'Aboriginal.' We, however, prefer to describe this population as 'Indigenous.'
2. It should be noted that while the census data were sufficient to allow us to disaggregate Indigenous from racialized faculty (reference to in the census as 'visible minority'), for our survey data, the numbers were too small for us to do the same. Hence, our reference here to racialized faculty is a combination of Indigenous and racialized respondents.
3. Due to limited resources and using an analytic framework that foregrounds race and racism, we were only able to get at social class in relation to tenured versus non-tenured faculty. For basically the same reasons, we were unable to include disability because that data are even more unavailable than for race.
4. Interestingly, the situation has remained about the same and slightly improved for Indigenous faculty.
5. Many of these respondents could have been graduates of a Canadian university, but many were likely foreign hires. This raises concerns about what is happening to Canadian-born racialized doctorates who appear not to be transitioning into the academic labor market. It is a problem seen in other job sectors, and one that is raising concern over potential inequality and alienation from Canadian society (Reitz and Bannerjee 2007).
6. In their study of Black male faculty in white university campuses, Griffin, Ward, and Phillips (2014, 1369) found that their everyday routine experiences not only led to 'microaggressions,' but also to psychological states such as 'imposter syndrome and racial battle fatigue. … Imposter syndrome refers to strong feelings of self-doubt despite one's intelligence and credentials … while racial battle fatigue marks the physical, mental, and emotional stress that racialized oppression brings forth.'

Disclosure statement

No potential conflict of interest was reported by the authors.

References

Abella, R. 1984. *Equality in Employment: A Royal Commission Report*. Ottawa: Minister of Supply and Services.

Ahmed, S. 2012. *On Being Included: Racism and Diversity in Institutional Life*. Durham: Duke University Press.

Bell Jr., D. A. 1980. "Brown V. Board of Education and the Interest-convergence Dilemma." *Harvard Law Review* 93: 518–533.

Braedley, S., and M. Luxton. 2010. "Competing Philosophies: Neoliberalism and the Challenges of Everyday Life." In *Neoliberalism and Everyday Life*, edited by S. Braedley and M. Luxton, 3–21. Montreal: McGill-Queen's University Press.

British Sociological Association Annual Conference. 2015. *Societies in Transition: Progression or Regression*. Glasgow: Caledonian University.

Carty, L. 1991. "Black Women in Academia." In *Unsettling Relations*, edited by Himani Bannerji, Linda Carty, Karl Dehli, Susan Heald, and Kate McKenna, 13–44. Toronto: Women's Press.

Crenshaw, K. 2002. "The First Decade: Critical Reflections, or 'a Foot in the Closing Door.'" *UCLA Law Review* 49: 1343–1372.

Crenshaw, K., N. Gotanda, G. Peller, and K. Thomas. 1995. *Critical Race Theory: Key Writings That Formed the Movement*. New York: New Press.

Delgado, R., and J. Stefancic. 2012. *Critical Race Theory: An Introduction*. 2nd ed. New York: New York University Press.

Drolet, D. 2009. "Universities Not Facing up to Race Issues, Say Scholars." *University Affairs*. Accessed March. http://www.universityaffairs.ca/news/news-article/universities-not-facing-up-to-race-issues/

Dua, Enakshi. 2009. "Evaluation of the Effectiveness of Anti-racist Policies in Canadian Universities: Issues of Implementation of Policies by Senior Administration." In *Racism in the Academy*, edited by Frances Henry and Carol Tator, 49–74. Calgary: University of Toronto Press.

Dua, Enakshi with Nael Bhanji. 2012. "Exploring the Potential of Data Collected under the Federal Contractors Programme to Construct a National Picture of Visible Minority and Aboriginal Faculty in Canadian Universities." *Canadian Ethnic Studies* 44 (1): 49–74.

Dua, Enakshi, and Bonita Lawrence. 2000. "Whose Canada is It? White Hegemony in University Classrooms, with Bonita Lawrence." *Atlantis* 25 (2). Spring.

Gillborn, D. 2015. "Intersectionality, Critical Race Theory, and the Primacy of Racism: Race, Class, Gender, and Disability in Education." *Qualitative Inquiry* 21 (3): 277–287.

Giroux, H. A. 2014. *Neoliberalism's War on Higher Education*. Toronto: Between the Lines.

Gotanda, N. 1991. Cited in: S. F. Vaught Racism. *Public Schooling and the Entrenchment of White Supremacy*. Albany, NY: State University of New York Press, p. 16.

Griffin, K. A., J. C. Bennett, and J. Harris. 2013. "Marginalizing Merit?: Gender Differences in Black Faculty D/Discourses on Tenure, Advancement, and Professional Success." *The Review of Higher Education* 36 (4): 489–512.

Griffin, R. A., L. Ward, and A. R. Phillips. 2014. "Still flies in Buttermilk: Black Male Faculty, Critical Race Theory, and Composite Counter Story telling." *International Journal of Qualitative Studies in Education* 27 (10): 1354–1375.

Henry, F., and C. Tator, eds. 2009. *Racism in the Canadian University: Demanding Social Justice, Inclusion, and Equity*. Toronto: University of Toronto Press.

Henry, F., and C. Tator. 2012. "Interviews with Racialized Faculty Members in Canadian Universities." *Canadian Ethnic Studies* 44 (1): 75–99.

James, M., and S. Valluvan. 2014. "Higher Education: A Market for Racism?" *Darkmatter in the Ruins of Imperial Culture*. Accessed April. http://www.darkmatter101.org/site/2014/04/25/higher-education-a-market-for-racism/

Jayakumar, U. M., T. C. Howard, W. R. Allen, and J. C. Han. 2009. "Racial Privilege in the Professoriate: An Exploration of Campus Climate, Retention, and Stratification." *The Journal of Higher Education* 80 (5): 538–563.

Kurasawa, F. 2002. "Which Barbarians at the Gates? From the Culture Wars to Market Orthodoxy in the North American Academy." *Canadian Review of Sociology and Anthropology* 39 (3): 323–347.

Ladson-Billings, G. 1998. "Just What is Critical Race Theory and What's It doing in a Nice Field like Education?" *International Journal of Qualitative Studies* 11 (1): 7–24.

Law Commission of Ontario. 2012. *Vulnerable Workers and Precarious Work: Final Report.* Accessed December 2014. http://www.lco-cdo.org/en/vulnerable-workers-final-report

Luther, R. E., E. Whitmore, and B. Moreau, eds. 2003. *Seen but Not Heard: Aboriginal Women and Women of Colour in the Academy. Feminist Voices #11.* Ottawa: Canadian Research Institute for the Advancement of Women.

Mahtani, M. 2004. "Mapping Race and Gender in the Academy: The Experiences of Women of Colour Faculty and Graduate Students in Britain, the US and Canada." *Journal of Geography in Higher Education* 28 (1): 91–99.

Monture, P. 2010. "Race, Gender and the University: Strategies for Survival." In *States of Race: Critical Race Feminism for the 21st Century*, edited by S. Razack, M. Smith, and S. Thobani. Toronto: Between the Lines Press.

Monture-Angus, P. 1995. *Thunder in My Soul: A Mohawk Woman Speaks.* Halifax: Fernwood.

Monture-Angus, P. 1998. "Standing against Canadian Law: Naming Omissions of Race, Culture and Gender." *Yearbook of New Zealand Jurisprudence* 2: 7–29.

Mukherjee, A. P. 1994. *Oppositional Aesthetics: Readings from a Hyphenated Space.* Toronto: TSAR Publications.

Nakhaie, M. R. 2004. "Who Controls Canadian Universities? Ethnoracial Origins of Canadian University Administrators and Faculty's Perception of Mistreatment." *Canadian Ethnic Studies* 26 (1): 19–46.

Nakhaie, M. R. 2007. "Universalism, Ascription and Academic Rank: Canadian Professors, 1987–2000." *Canadian Review of Sociology and Anthropology* 44 (3): 361–386.

Newson, J. 2012. "University-on-the-Ground: Reflections on the Canadian Experience." In *Reconsidering Knowledge: Feminism and the Academy*, edited by M. Luxton and M. J. Mossman, 96–127. Halifax: Fernwood Publishing.

Prentice, S. 2000. "The Conceptual Politics of Chilly Climate Controversies." *Gender and Education* 12: 195–207.

Razack, S. 1998. *Looking White People in the Eye: Gender, Race and Culture in Courtrooms and Classrooms.* Toronto: University of Toronto Press.

Reitz, J., and R. Bannerjee. 2007. "Psychosocial Integration of Second and Third Generation Racialized Youth in Canada." *Canadian Diversity/Diversité Canadienne* 6 (2): 54–57.

Smith, M. S. 2010. "Gender, Whiteness, and 'Other Others' in the Academy." In *States of Race: Critical Race Feminism for the 21st Century*, edited by S. Razack, S. Thobani, and M. Smith. Toronto: Between the Lines Press.

Thomas, M. 2010. "Neoliberalism, Racialization, and the Regulation of Employment Standards." In *Neoliberalism and Everyday Life*, edited by S. Braedley and M. Luxton, 68–89. Montreal: McGill-Queen's University Press.

Thornton, M. 2012. "Universities Upside down: The Impact of the New Knowledge Economy." In *Reconsidering Knowledge: Feminism and the Academy*, edited by M. Luxton and M. J. Mossman, 76–95. Fernwood: Halifax.

Williams, Patricia J. 1992. *The Alchemy of Race and Rights: A Diary of a Law Professor.* Cambridge, MA: Harvard University Press.

What style of leadership is best suited to direct organizational change to fuel institutional diversity in higher education?

Ryan P. Adserias, LaVar J. Charleston and Jerlando F. L. Jackson

ABSTRACT

Implementing diversity agendas within decentralized, loosely coupled, and change-resistant institutions such as colleges and universities is a global challenge. A shift in the organizational climate and culture is imperative to produce the change needed in order for a diversity agenda to thrive. Higher education scholars have consistently identified leadership styles as being among the chief contributing factors to successful institutional change, especially as it relates to diversity agenda efforts. This chapter first reviews the literature on forms of diversity agenda, paradigms of change and leadership style and then synthesizes results from 10 cases on proven strategies and offers implications on how different leadership styles can be applied to fuel institutional diversity efforts.

Introduction

As institutions embedded within broader society, colleges and universities are neither immune to the persistent challenges, nor to the rewards of promoting the values of social diversity, equity, and inclusion. In American higher education, the promotion of these values has not come without resistance, and institutions will continue to face significant internal and external challenges to the project of incorporating diversity into their organizational structures and cultures (Aguirre and Martinez 2006; Williams 2013). In response, scholars and practitioners alike have coalesced around the idea that higher education must undergo transformational change in order to reflect shifting demographic trends, to prepare students for an increasingly globalized economy and diverse workforce, and to embody the values of social and cultural pluralism and equity (Aguirre and Martinez 2006; Chun and Evans 2009; Williams 2013). These values have been broadly conceived of as 'the diversity agenda'.

Three areas of inquiry informed this study and are reviewed below. First, the relevant literature concerning the strategies guiding institutional diversity efforts, or the diversity agenda, is reviewed. Next, an overview of two paradigms of change implicated in the scholarship concerning the diversity agenda – namely co-optative and transformative change – is presented. Finally, the topic of leadership is explored with particular attention paid to three leadership style paradigms germane to this topic: transactional, transformational, and full-range leadership.

Diversity agendas

As outlined above, there are several broad areas of challenge and opportunity diversity agendas, described by Williams (2013) as: the social justice rationale, the educational benefits rationale, and the business rationale. The social justice rationale refers to the imperative that higher educational institutions become more reflective of shifting demographic trends, and address both past and present identity-based social inequities. The educational benefits rationale is grounded in research findings demonstrating the benefits of attracting and retaining students from diverse backgrounds to institutions' pedagogical and human development missions. The business rationale refers to the need for institutions to grow more inclusive to better compete in the market for talented students, faculty, and staff and to prepare students for an increasingly globalized economy and diverse workforce (Williams 2013).

In response to these areas of challenge and opportunity, American colleges and universities have increasingly come to recognize diversity as a matter of strategic importance (Williams 2013). Many institutions have developed institutional policy statements, broadly conceived of as diversity agendas, as a means of signaling commitment and organizing their diversity-related strategies (Anderson 2008; Iverson 2007, 2008; Kezar and Eckel 2008; Kezar, Eckel et al. 2008; Williams 2013). The term 'diversity agenda' is loosely defined. Indeed 'there is no consensus across institutional types concerning an agreed-upon format for a campus diversity plan' (Anderson 2008, 38). Kezar and Eckel (2008, 401) offered that some institutions refer to the diversity agenda as encompassing 'efforts to change the campus to be more inclusive', and that '[a] diversity agenda or initiative is multifaceted and attempts to integrate diversity into the structure, culture, and fabric of the institution so that it is truly institutionalized'. Other scholars (Anderson 2008; Iverson 2007, 2008; Williams 2013) defined offered greater specificity around the elements of a diversity agenda. Iverson (2008, 183) for example suggested that diversity agendas (or action plans) are 'policy documents [that] are a primary means by which universities advance recommendations regarding their professed commitment to equal access and an inclusive environment for all members of the campus community'. Williams (2013, 303) defined diversity plans as, 'any intentionally created document that includes a diversity definition, rationale, goals, recommended actions, assignments or responsibility, timelines, accountability processes, and a budget'. Regardless of name or degree of specificity, diversity agendas ideally function as a blueprint for action, as a guide for members of a campus community and leaders tasked with implementing its recommendations for change (Williams 2013) and more importantly, as a signal of institution's desire to change its culture to become an inclusive organization (Aguirre and Martinez 2006).

Despite their symbolic significance as institutional policy and strategy artifacts, diversity agendas have high failure rates (Williams, Berger, and McClendon 2005). Among the reasons for failure described by Williams, Berger, and McClendon (2005) are two connected issues germane to this discussion: namely, their low rates of success in changing organizational culture, and absent, weak, or insufficient campus leadership. In the following section, we briefly examine paradigms of change applied to diversity agendas and leadership styles, and their relationships to organizational change.

Paradigms of change

The degree to which organizational change occurs has been characterized as being either first- or second-order. First-order change accounts for 'those minor improvements and adjustments that do not change the system's core, and occurs as the system naturally grows and develops' (Levy and Merry 1986, 5). Second-order change represents a fundamental alteration of an organization's 'underlying values or mission, culture, functioning processes, and structure of the organization' (Kezar 2001, 16). Scholars have identified co-optative and transformational change as two types of change paradigms employed by postsecondary institutions in response to diversity, and have characterized them as representing 'competing organizational strategies' (Aguirre and Martinez 2006, 48).

Co-optative change strategies are a first-order change, and refer to superficial efforts that '[absorb] new elements into the leadership or policy determining structure of an organization as a means of averting threats to its stability or existence' (Selznick 1948, 34). Aguirre and Martinez (2006, 56) noted that co-optation refers to 'strategies to address discrimination and social justice issues linked to diversity'. Co-optative change is furthermore characterized as a rational-bureaucratic approach, that is managerial in nature, planned low-level change (Aguirre and Martinez 2006). Co-optation is counter to the aims of the diversity agenda as co-optation 'use[s] diversity dimensions in the organizational culture – minority faculty, multiculturalism in the curriculum, and role models for minority students – as buffers to protect organizational culture rather than to change it' (Aguirre and Martinez 2006, 56).

In contrast, Aguirre and Martinez (2006) characterized strategies aimed toward engendering deeper, cultural change as transformational change. Emanating from their half-decade long study of transformation, Eckel and Kezar (2003, 27) defined transformational change 'as affecting institutional cultures, as deep and pervasive, as intentional, and as occurring over time'. Transformational approaches to organizational change are planned, second-order responses to shifts in the external environment aimed toward changing organizational culture 'without necessarily the whole organization' (Aguirre and Martinez 2006, 56). In the context of diversity agendas, transformational change aims to interweave principles of social justice and multiculturalism into the fabric of the organization including its culture (Aguirre and Martinez 2002, 2006; Williams 2013). Organizational change scientists have observed that transformational change is difficult to achieve and sustain owing, in part, to organizations' complexity. Examples of successful transformational change in higher education are scant and in those organizations that experienced transformation, the changes did not meet expectations (Kezar 2001). Transformational change, including transformational change related to diversity, is possible (Kezar and Eckel 2008; Williams 2013). Here, leaders are expected to outline a shared vision of the organization's future, and to facilitate the necessary structures and processes through which organizational members engage in learning (Eckel and Kezar 2003; Kezar 2005).

In order to achieve the fundamental goals set by the diversity agenda, higher educational institutions must undergo transformational change and the target of transformation must be organizational culture (Aguirre and Martinez 2002, 2006; Williams 2013). Cultural change is a process that scholars believe necessitates strong leadership (Kezar 2001).

Leadership style

There is no consensus around what qualifies as 'the best style' of leadership (Bolman and Deal 2008). As it pertains to the transformation of organizational culture prescribed by diversity agendas in higher education, some scholars (e.g. Aguirre and Martinez 2006) have suggested transformational leadership styles are particularly effective. Other scholars have observed that presidents report utilizing an amalgam of styles to lead the implementation of diversity agendas (transformational, transactional, and laissez-faire – or full-range leadership), (Kezar and Eckel 2008). Still others offered that effective diversity leaders master a panoply of leadership styles inclusive of Bolman and Deal's (2008) structural, political, and symbolic frames, and Birnbaum's (1988) collegial leadership frame (Williams 2013; Williams and Wade-Golden 2013). In this review, we use 'transactional leadership' and 'transformational leadership' as two bookend benchmarks on the scale of leadership styles in order to analyze and describe the characteristics of other styles that fall in between.

Transactional leadership

Bensimon, Neumann, and Birnbaum (1989, 10) described transactional leadership as 'a relationship between leaders and followers based on an exchange of valued things, which could be economic, political, or psychological in nature'. Transactional leaders build relationships based primarily on the principles of trust and honesty and in the service of maintaining organizational order and culture (Bensimon et al. 1989). Transactional leaders typically work with their subordinates in a reciprocal manner to maintain the organizational status quo. Bass and Riggio (2006) described transactional leaders as utilizing: (a) contingent rewards (characterizes leaders who promise and deliver rewards based on satisfactory performance); (b) passive management by exception (characterizes leaders who await a subordinate's mistakes to take corrective action); and (c) active management by exception (characterizes leaders who point out subordinate's mistakes in order to warn others), as a means of exerting power and influence. The managerial nature of transactional leadership is evident in characteristics of the style outlined by Bass and Riggio (2006), and is evinced by the emphasis placed on setting expectations, monitoring and rewarding compliance and progress, and punishment or correction of deviation.

Owing to its emphasis on maintaining organizational cultures, Bensimon and colleagues (1989) suggested college and university leaders are most likely to employ a transactional style of leadership. They argued that a transactional style is most applicable in higher education settings where leadership is more diffuse, and values more likely to be localized than broad and organization-wide (Bensimon et al. 1989). Furthermore, this leadership style most accurately describes the nature of how 'college and university presidents can accumulate and exert power by controlling access to information, controlling the budgetary process, allocating resources to preferred projects, and assessing major faculty and administrative appointments' (Bensimon et al. 1989, 39). However, Bass (1990, 21), based on research of private sector organizations, noted that this exchange-based style of leadership is 'ineffective and, in the long run, may be counterproductive', and furthermore that the efficacy of a transactional style of leadership is dependent upon a leader's access and ability to distribute rewards and, 'whether the employees want the rewards or fear the penalties'.

Transformational leadership

Transformational leadership has been identified as having greater potential for leading the type of large-scale, long-term organizational, and cultural changes necessitated by the diversity agenda (Aguirre and Martinez 2002, 2006). Burns (1978, 20) described the transformational style as moral leadership that 'raises the level of human conduct and ethical aspiration of both leader and led, and thus it has a transforming effect on both'. Successful transformational leaders serve as teachers and moral guides, and 'interpret [organizational] culture and manipulate symbols', as a means of achieving organizational change (Tierney 1989, 160). Transformational leaders employ the following tactics: (a) idealized influence (leading by example); (b) inspirational motivation (leaders inspire commitment by challenging subordinates, providing meaning to their work, and delineating an attractive vision of the future); (c) intellectual stimulation (leaders inspire and support creative thinking and problem solving); and (d) individualized consideration (leaders provide coaching and mentoring to subordinates in order to help achieve both personal and collective goals and growth) (Bass and Riggio 2006).

Transformational approaches to leadership hold potential for both understanding and conceptualizing the transformative changes necessary to ameliorate systemic oppressions, such as those based on race, ethnicity, gender, and other identities that are socially marginalized (Aguirre and Martinez 2002, 2006; Tierney 1989). Additionally, the transformational style may be well suited to leading the sense-making and organizational learning processes necessary for transformational change to occur (Kezar 2001, 2014). Despite its promise, some authors have suggested transformational leadership as futile and may encounter significant resistance. As Bensimon, Neumann, and Birnbaum (1989, 72) noted, transformational style 'in many ways may not be compatible with the ethos, values, and organizational features of colleges and universities'. Indeed, in higher education institutions, researchers have found a mixture of both transformational and transactional, or a full-range leadership style, to be both effective and most likely to be used by leaders.

Full-range leadership

Transactional and transformational leadership may be viewed as occupying two polar ends of the power and influence leadership paradigm, while a full-range leadership style represents the midpoint between the two approaches. The origins of full range leadership are traced to Bernard Bass' work integrating the '*charismatic-transforming-leadership* approach', [and the] '*bureaucratic-transactional-management* approach' (Antonakis and House 2002, 7; italics original). Full-range leadership combines aspects of both transformational and transactional leadership styles. Indeed, Bass and Riggio (2006, 9) noted that the full-range leadership model *requires* leaders to '[display] each style to some amount'. Full range leadership is comprised of the characteristics of transformational leadership (i.e. idealized influence, inspirational motivation, intellectual stimulation, and individualized consideration), and transactional leadership (i.e. contingent rewards, passive management by exception, active management by exception).

Leadership frames

Bolman and Deal's (2008) four frame theories of organizations and leadership are characterized as representing the: (a) structural frame – the use, coordination, and control of formal
organizational structures and hierarchies as a means of advancing organizational priorities
and achieving its goals; (b) human resource frame – organizational functioning, efficacy, and
success as contingent upon meeting the needs, and appealing to interests of its members;
(c) political frame – utilizes organizational power inequities, and competing factions and
constituencies that are both internal and external to the organization to mobilize a leader's
agenda; and (d) symbolic frame – relies on a leader's communicative prowess, appeal to
shared values and morals, and the use of powerful organizational symbolism. In addition to
Bolman and Deal's (2008) four-frame theory, Birnbaum's (1988) *collegial frame* is appropriate
for the loosely coupled type of organization characterizing colleges and universities (Weick
1976), where power is shared through various academic units and the structures of shared
governance. In the *collegial frame* leaders leverage the institution's established structures of
shared governance, and adhere to cultural norms of academe and the institution as a means
of deriving consensus among organizational power factions and constituencies (Birnbaum
1988). Aligning the various frames of Bolman and Deal's (2008) theory, and Birnbaum's
(1988) collegial frame with the broader transformational and transactional leadership style
paradigms, we conceived of the structural and political frames representative of the transactional, and the human resources, symbolic, and collegial frames as representative of the
transformational styles of leadership. The structural and political frames and transactional
leadership paradigm emphasize hierarchical structures, strategic planning, and the use of
incentives, rewards, and punishments as leadership technologies. In contrast, the human
resources, symbolic, and collegial frames and transformational leadership paradigm all
emphasize the need for leaders to appeal to their followers' sense of morality, to themselves
maintain and exhibit a high moral standard, to communicate a strong organizational vision,
and to engage, empower, and inspire their followers.

Leaders make choices about which style of leadership to employ based on a complex
calculus accounting for many factors including the dimensions of the challenges facing the
institution (Bensimon 1993), their social identity (Kezar and Eckel 2008), location within
the organization (Brown and Moshavi 2002), and organizational culture (Kezar, Carducci,
and Contreras-McGavin 2006), among other factors. As we have demonstrated earlier,
higher education scholars and practitioners have, in general, reached consensus identifying
transformational, as opposed to co-optative change, as being optimum for advancing an
institutional diversity agenda. Less clear, however, is which style of leadership is best suited
to leading these diversity-oriented change agendas. To that end, the question guiding our
literature-based research was: which style of leadership is best suited to implementing the
diversity agenda in institutions of higher education?

Method

Our aim is to identify the optimal style(s) of leadership for implementing the diversity
agenda in colleges and universities and to 'summarize the accumulated state of knowledge
concerning the relation(s) of interest and to highlight important issues that research has
left unresolved' (Cooper 1982, 292). We approached our review of the scholarly literature
following four steps. First, the authors read three foundational monographs to help guide

both our theoretical understanding and search procedure: Aguirre and Martinez's (2006) *Diversity Leadership in Higher Education*, Williams (2013) *Strategic Diversity Leadership: Activating Change and Transformation in Higher Education*, and Williams' and Wade-Golden's (2013) *The Chief Diversity Officer: Strategy, Structure, and Change Management*. Arising from our reading, we identified organizational and cultural change as central themes to help guide our later search of scholarly literature. Next, we conducted an extensive search of five research databases. Our search yielded a total of 448 sources. Third, each of the 448 sources was reviewed to determine their relevance to the research question. Almost all of the manuscripts were deemed to be irrelevant to the research question, with the exception of three manuscripts. In the fourth and final step, an ancestry approach was employed to identify additional sources cited in both the foundational texts and those discovered through database searches.

Search procedure

Aided by the theoretical concepts identified in our reading of the three foundational texts, we conducted an extensive search of five electronic scholarly research databases: (a) *Academic Search Premier*; (b) *ERIC*; (c) *Education Research Complete*; (d) *Education Full Text*; and (e) *JSTOR*. For each database we conducted two queries using Boolean search techniques, and restricted our search to peer-reviewed publications using the following terms: (a) *organizational change OR culture change* and *leadership* and *diversity* and *higher education*; and (b) *leadership* and *diversity* and *higher education*. Each database entry was imported into the open source citation management program *Zotero*, which captured the bibliographic metadata, and when available, the abstract for each article. After eliminating duplicates, this step yielded 448 total results. Next, the abstract for each result was reviewed and 149 manuscripts were determined to be relevant. Each of the 149 texts was reviewed to determine whether the concept of leadership was addressed as a factor in diversity on college campuses; of the 149 reviewed, only 6 manuscripts were relevant. Of the texts deemed to be irrelevant ($n = 143$), most implicated leadership only insofar as individual leaders were involved in the design or implementation of diversity efforts and initiatives, but did not sufficiently discuss the manner and degree to which leaders were involved and were thus deemed irrelevant to this inquiry. Fourth and finally, an ancestry search procedure (Cooper 1982) was employed. The ancestry approach allows researchers to identify sources cited by previous authors that are overlooked. Using this approach, one additional text was identified.

Analysis procedure

A total of 10 manuscripts met the criteria for inclusion in our analysis, and are listed in Table 1, along with the mode of discovery. In order to achieve our goal of discerning which style or styles of leadership are best suited to implementing the diversity agenda in institutions of higher education, our analysis paid particular attention to two areas: first, we examined how each manuscript addressed the type and role of diversity-related organizational change; and second, how authors approached the concept of leadership and leadership style. Although the primary concern of this chapter is leadership style, separating leadership from organizational change paints an incomplete picture as the different paradigms of diversity-related change are closely associated with particular approaches to leadership (Aguirre

Table 1. Sources analyzed and mode of discovery.

Manuscript author(s) and year	Mode of discovery
Aguirre and Martinez (2006)	Foundational
Anderson (2008)	Ancestry
Chun and Evans (2009)	Search
Kezar, Eckel et al. (2008)	Search
Kezar and Eckel (2008)	Search
Kezar, Glenn et al. (2008)	Search
Kezar (2007)	Search
Kezar (2008)	Search
Williams and Wade-Golden (2013)	Foundational
Williams (2013)	Foundational

and Martinez 2006). As we observed early on in this research, few scholars addressed or referred to leadership style outright, and only one manuscript (Kezar and Eckel 2008) referred to transactional, transformational, or full-range leadership styles explicitly. For our analysis procedure, we first read each of the 10 manuscripts and developed broad inductive codes for references to leadership, and styles of leadership based on our understanding of the literature. Next, we employed a deductive strategy to determine whether leadership strategies, tactics, and style aligned with the theoretical literature. In a final step, we also employed inductive techniques (Boyatzis 1998) to identify important themes that related to leadership, but were not necessarily addressed in the transactional, transformational, or full-range leadership literatures.

Literature synthesis

Among the 10 manuscripts we analyzed for this study, two major themes concerning the type of change and style of leadership emerged. First, each of the manuscripts identified transformational change as the type of change necessary for addressing campus diversity, and that organizational culture should be the target of change. Many manuscripts – though not all – conceived of the transformation of organizational culture as occurring in stages. Second, most authors emphasized the importance of leaders in guiding the transformation of organizational culture, and that leaders employed a number of leadership styles and strategies that were either transformational or transactional, or both (i.e. full range leadership). Our results led us to conclude that there is not one style of leadership best suited to directing organizational change for diversity in higher education. Rather, much of the literature reviewed for this article is in agreement with Williams (2013, 206) assertion that 'campus diversity champions must be sophisticated in their approach and willing to work against the time-honored traditions and time-bound bureaucracies that render academic institutions so resistant to change'. Our findings also led us to concur with Kezar, Eckel and associates' (2008, 87) assertion 'that leaders need to adopt a complex and varied approach', to leadership inclusive of both transactional and transformational styles. The style most closely aligned with the level of sophistication, complexity, and variation necessary to lead the organizational change necessitated by the diversity agenda is the full-range leadership approach. Only Kezar and Eckel (2008) however, expressly cited this style as being employed by leaders pursuing organizational change related to diversity. Finally, some authors avoided characterizing leadership style in terms of being either transformational or transactional, and opted instead to characterize leadership style in terms of Bolman and Deal's (2008) four-frames theory of organizations and leadership. In addition to the four frames, Williams (2013) and Williams and Wade-Golden (2013) included the 'collegial leadership frame'. In the next session, we discuss the type and target of change these leaders have on their agenda.

Target and phases of change

Transformational change was identified either explicitly (i.e. Aguirre and Martinez 2006; Anderson 2008; Chun and Evans 2009; Williams 2013; Williams and Wade-Golden 2013), or implicitly (i.e. Kezar 2007, 2008; Kezar and Eckel 2008; Kezar, Eckel et al. 2008; Kezar, Glenn et al. 2008), as necessary for the organizational change prescribed by the diversity agenda. Of the sources that implicitly identified transformational change, all discussed initiatives or efforts aimed toward engendering deep, pervasive, and long-lasting changes that align with the definition of transformational change (Eckel and Kezar 2003). Organizational culture change was seen as necessary for addressing diversity in colleges and universities in order for institutions to become inclusive (Aguirre and Martinez 2006).

Implementation of the diversity agenda and changing organizational culture occurs in phases or stages, and leaders described employing differing leadership approaches and styles concomitant with various phases or stages (Anderson 2008; Kezar 2007, 2008; Kezar and Eckel 2008; Kezar, Eckel et al. 2008; Williams 2013). In one phase model, Kezar (2007) offered that the institutionalization of diversity initiatives or agendas occur in three phases; the first, the structural phase, refers to campuses that typically lack a diversity agenda and few conversations about diversity occur. The second, the behavioral phase, refers to institutions that 'have a diversity agenda and on-going conversations related to race, gender, social class, and other aspects of diversity ... [and] have a clear rhetoric related to diversity and supporters committed to diversity' (Kezar 2007, 418). Finally campuses in the third, cultural phase '[h]ave mostly institutionalized a diversity agenda ... and have regular monitoring mechanisms to keep track of their diversity efforts and ensure they are making progress' (Kezar 2007, 418–419). Institutions in the third or cultural phase have achieved the cultural transformation necessitated by the diversity agenda.

Williams (2013) offered a slightly different model that posits institutions move through four stages of institutionalizing diversity. First, the start-up stage campuses as those where 'diversity is either not on the radar or is considered a distraction from advancing the goals of academic excellence ... diversity is not defined as an institutional priority' (Williams 2013, 197), where there are few initiatives and scant infrastructures for supporting diverse students, faculty, and staff. The second, transitional stage concerns institutions where 'the diversity discussion begins to emerge as a point of conversation among senior leadership ... no cohesive institutional framework or agenda has emerged' (Williams 2013, 199–200). In stage two campuses extant diversity efforts are typically isolated. As institutions progress to the third, mature implementation stage the diversity agenda begins to emerge as a priority, and '[s]enior leaders generally have a strong awareness of diversity issues ... they still do not define their work as inspiring institutional change and transformation, but rather in more incremental terms' (Williams 2013, 201). Stage three campuses 'may spend a lot of time in the mature implementation phase ... [and many] have developed multiple diversity plans' (Williams 2013, 201–202), though these plans achieve limited success due to the lack of a robust accountability regime. The fourth and final inclusive excellence stage refers to institutions where '*diversity* is defined broadly and exists at the highest level of institutional importance as foundational to mission fulfillment and academic excellence … [and] has become a cultural value that manifests itself in myriad ways' (Williams 2013, 203; italic in original). In other words, campuses having achieved the fourth of Williams' stages have transformed themselves so that '[b]eyond dedicated diversity roles, the broader campus community and leadership play an active role in diversity efforts … [d]iversity matters are substantively integrated into the curriculum and cocurriculum' (Williams 2013, 203).

Senior-level leadership

Aguirre and Martinez (2006), Chun and Evans (2009), and Williams (2013) implicate an array of senior leaders including 'presidents, provosts, deans, and other administrators' (Aguirre and Martinez 2006, 81). Primary attention is paid to presidential leadership by Kezar, Eckel et al. (2008), Kezar and Eckel (2008) and Kezar (2007, 2008). Although this narrow focus is justified by presidents' unique abilities to muster and deploy the necessary resources for transformational change, the attention paid to presidents in these texts owes as much credit to the fact that each manuscript arose from a long-term study of presidential leadership of transformational change. The importance of senior-level leaders' commitment was underscored by Williams (2013, 176) thus: '[m]ore than any other factor, the leadership's commitment to deep and meaningful change will determine whether the institution builds capacity for the long-term'. If the diversity agenda is to achieve its goal of a deep and lasting transformation of organizational culture, senior leaders such as 'presidents, provosts, deans, and other administrators' (Aguirre and Martinez 2006, 81), must be engaged.

Examples of other senior-level leaders were offered by Williams and Wade-Golden (2013) and Williams (2013) in their study of the chief diversity officer (CDO) position. Although Chun and Evans (2009, 50) identified the CDO 'as a structural best practice', cohesion around a definition and the scope of the CDO's role is lacking, Williams and Wade-Golden (2013, 32) state:

> The CDO is a boundary-spanning senior administrative role that prioritizes diversity-themed organizational change as a shared priority at the highest levels of leadership and governance. Reporting to the president, provost, or both, the CDO is an institution's highest-ranking diversity administrator. The CDO is an integrative role that coordinates, leads, enhances, and in some instances supervises formal diversity capabilities of the institution in an effort to create an environment that is inclusive and excellent for all.

In addition to senior-level leaders, other institutional leaders are important players in advancing the diversity agenda and driving the transformation of organizational culture. For example, Anderson (2008) noted the importance of faculty leadership in infusing diversity into the teaching and learning enterprise. Chun and Evans (2009) observed that in order for diversity to become fully integrated into institutions of higher learning, members of campus community must be empowered to play a leadership role, for leadership to be distributed, and decision-making to be democratic. They (Chun and Evans 2009, 44) implicated a variety of organizational members including senior-level leaders (e.g. boards of trustees, presidents, and CDOs), academic and other administrative leaders, and other 'campus constituent groups'.

Table 2 shows which leadership styles, if any, texts cited, or whether we inferred a style from our analysis of the texts. In two cases, the transformational style was explicitly implicated as part of the foundation of a particular leadership strategy, such as diversity leadership

Table 2. Described leadership style, implicitly or explicitly implicated.

Source	Described leadership style	Explicitly or implicitly implicated
Aguirre and Martinez (2006)	Diversity leadership	Transformational leadership explicitly implicated
Anderson (2008)	Transformational leadership	
Chun and Evans (2009)	Reciprocal empowerment; diversity leadership	Transformational leadership implicitly implicated
Kezar and Eckel (2008)	Full-range leadership	Full-range leadership explicitly implicated
Kezar, Eckel et al. (2008)	Four frames	Full-range leadership implicitly implicated
Kezar, Glenn et al. (2008)	None	
Kezar (2007)	None	
Kezar (2008)	Political frame	
Williams and Wade-Golden (2013)	Strategic diversity leadership	
Williams (2013)	Strategic diversity leadership	

(Aguirre and Martinez 2006; Chun and Evans 2009). Transformational leadership is implicitly implicated as one aspect of the full-range style in just over half of the texts reviewed, and was in one case explicitly implicated as part of a full-range leadership approach (Kezar and Eckel 2008). In half of the texts (Kezar 2007, 2008; Kezar, Eckel et al. 2008; Williams 2013; Williams and Wade-Golden 2013), the full-range style leadership was implied. Only Anderson (2008) explicitly implicated the transformational style solely, and none of the literature implicated transactional leadership in isolation.

Although none of the literature suggested approaching diversity-related cultural change via transactional leadership approaches alone, they advanced transactional approaches. For example, among the studies concerning presidential leadership and Kezar's (2007) phase-based approach to implementing the diversity agenda, presidents employed transactional approaches to strategic planning. Kezar (2007) observed that in each phase of the diversity agenda implementation process, presidents used transactional approaches such as linking the diversity agenda to resource allocation decisions as a means of implementing incentive and reward structures, and instituting penalties for failing to advance diversity-related goals. Kezar, Eckel et al.'s (2008, 88) study suggested that presidents employ a transactional approach to distribute resources and incentivize participation and adoption of the agenda at 'the beginning of campus efforts', and in later phases, incentives 'became important to foster student retention and [to] reward units for meeting key goals'. The use of data also featured prominently in some authors' (Chun and Evans 2009; Kezar 2007; Kezar and Eckel 2008; Kezar, Eckel et al. 2008; Williams 2013; Williams and Wade-Golden 2013) discussions of accountability mechanisms which are aligned with transactional leadership approaches. Throughout the texts that discussed data-based accountability regimes, the management by exception aspect of transactional leadership was common. Kezar (2007) for example, found that accountability was achieved using data to inform decision-making processes, presumably to mete out punishments and rewards. Other examples of management by exception included using data to hold deans accountable for meeting goals related to faculty diversification initiatives (Kezar, Eckel et al. 2008).

Aguirre and Martinez (2006) and Anderson (2008) cited the transformational style as best suited to advancing the diversity agenda, and Chun and Evans (2009) advanced a leadership strategy that employed a transformational style. Aguirre and Martinez (2006, 36) emphasized the transformational style in their 'diversity leadership' model, and noted that 'transformational leadership is conceived as a type of leadership needed by organizations to respond and adapt to environmental change, that is, demographic and cultural diversity'. Moreover Aguirre and Martinez (2006, 36) emphasized that 'transformational leadership enables the organization to be seen as responding to the collective need for identity and commitment between persons and organizational culture'. Further underscoring the importance of the transformational approach, Anderson (2008, 5) suggested that diversity is '... best served by ... leaders who understand the forces that pressure organizations to change and who exhibit the characteristics associated with transformational leadership' (5). Chun and Evans (2009) suggested that a reciprocal empowerment approach was well suited to implementing diversity-related change. Similar to transformational leadership, reciprocal empowerment appeals to organization members' values and morals, and both approaches empower individuals and create supportive and inclusive environments as a means of leading change (Chun and Evans 2009; Kezar and Eckel 2008).

The transformational style of leadership was implicated in most of the 10 texts we analyzed, and most often referred to leaders' ability to create and maintain relationships, create and communicate a vision of an inclusive and equitable campus, and to inspire action in others. Kezar, Eckel et al. (2008) found that the presidents cultivated relationships, or a 'web of support', throughout the campus and wider community as a means of implementing their agendas. Creating and communicating a vision also emerged as a core transformational leadership strategy. Kezar, Glenn et al. (2008) found that one successful president engaged the entire campus in creating a vision related to diversity. Engaging the campus in developing a collective vision evokes several aspects of the transformational leadership style including inspirational motivation, intellectual stimulation, and individualized consideration. This approach empowered institutional members to think critically and creatively about themselves, their work, and the institution, while creating a supportive environment for individuals' and groups' thoughts, feelings, and experiences to be heard.

Although Kezar and Eckel (2008) alone explicitly cited full-range leadership as best suited to leading the organizational change necessitated by the diversity agenda, a combination of transactional and transformational styles were implicated in the majority of texts analyzed for this chapter. A full-range style of leadership was implicated by Kezar, Eckel, and colleagues' (2008) use of the four frames theory of leadership, and is implicit in Williams (2013) and Williams and Wade-Golden's (2013) use of the 'strategic diversity leadership' concept. Full-range leadership is implicit in Kezar's (2008) study of political leadership, and in her 2007 research concerning phases of the institutionalization of the diversity agenda. Lastly, full-range leadership was also implied in Kezar, Eckel et al.'s (2008) study concerning the *Diversity Scorecard Project*, though to a lesser degree than was evident in other studies. Presidents' deployment of transactional and transformational styles was found by Kezar and Eckel (2008) to be dependent both on contextual and situational factors. They concluded 'presidents appear to have examined the institutional culture in order to determine whether transactional or transformational leadership will better align with campus expectations' (Kezar and Eckel 2008, 399). This finding is echoed in Kezar (2008), which suggests that presidents routinely surveyed their institutions, what Bolman and Deal (2008) in their political frame called 'mapping the political terrain', to determine proactive strategies for ameliorating resistance to the diversity agenda, and whether and which allies and coalitions to engage, and how.

The full-range style was described among those texts that conceived of the diversity agenda as being implemented in either of the two phases and stage models described earlier.

As institutions cycle through three phases or four stages of diversity agenda implementation, presidents and diversity leaders choose among transactional and transformational approaches. For example, following Kezar's (2007) three-phase model, Kezar and Eckel (2008) found that the presidents of phase one institutions reported using transformational approaches such as idealized consideration to engage in institution-wide dialog and to role model new institutional values; and employed idealized influence to communicate a vision of a diverse and inclusive campus. Transactional strategies were most likely to be employed by leaders of middle phase or stage institutions, as they sought to 'broaden the ownership' (Kezar 2007) of the agenda, and began to use data to drive decision-making processes and implement accountability measures. For example, Kezar (2008) noted that faculty diversification efforts were among the most likely initiatives to spur faculty resistance to the diversity agenda, and necessitated transactional approaches such as utilizing data to undermine arguments against hiring diverse faculty, creating financial rewards and incentives (Williams 2013), and employing strategies to hold academic leaders accountable (Kezar, Eckel et al. 2008).

Transformational styles emerged as institutions progressed to the middle phase or stage of implementing the diversity agenda. Kezar (2007), Kezar and Eckel (2008), Kezar, Eckel, and colleagues (2008) and Williams (2013) noted the importance of 'revitalizing the agenda' by employing transformational strategies to appeal to individuals' and groups' experiences, ideas, and creativity. In later stages of the diversity agenda implementation process, leaders continued to employ both transactional and transformational strategies. On phase three campuses, Kezar (2007, 431) observed that presidents employed transformational leadership approaches to

> helping members of campus and the community surrounding the campus learn about diversity in various forms … [and] focused on challenging traditional values and ways of doing work and on creating ways that are more supportive of students from diverse backgrounds. (431)

Later phase campuses also invested time and energy in developing the capacity and learning of its members through organizational learning processes and faculty mentoring programs (Kezar 2007; Kezar, Eckel et al. 2008). As it pertains to transactional strategies, most texts referred to the accountability regimes of later-stage institutions as being among the most robust and wide spread (Kezar 2007; Kezar and Eckel 2008; Kezar, Eckel et al. 2008; Williams 2013). This observation seems largely a priori however, as Kezar and Eckel (2008) noted there is a paucity of exemplar accountability regimes related to diversity. Finally, among the texts, transactional styles were identified as being most effective as institutions began to implement more difficult aspects of the diversity agenda, counted among which, not coincidentally, are those most necessary for facilitating cultural transformation (e.g. Kezar and Eckel 2008).

Conclusions and implications

Our analysis of the research examining the role of leadership style in implementing the diversity agenda in higher education led us to two conclusions. First, there is no singular style of leadership best suited to leading the implementation of the diversity agenda in colleges and universities. Rather, leaders employ both the transactional and transformational leadership styles in a manner closely resembling full-range leadership (Bass and Riggio 2006; Kezar and Eckel 2008). Second, we found that successful leaders pay close attention to various contextual features when determining whether to employ either transactional or transformational approaches. This analysis of the research literature underscores the scant research linking leadership style and organizational contextual features such as phase or stage of diversity agenda implementation.

These findings have several implications for both future inquiry and practice. Much of the research concerning leadership styles is more than a decade old and is in need of updating in the face of renewed interest in diversity leadership and recent protest movements over campuses' approaches to diversity. New thinking around the strategies addressing persistent and increasingly visible societal and institutional inequities is warranted. Research concerning the role of leadership in implementing the diversity agenda necessitates greater attention be paid to leaders other than presidents and CDOs. Among the manuscripts we analyzed for this chapter, only Anderson (2008) addressed the role of faculty leadership in a substantial way. While conventional wisdom, current practices, and our analysis implicate the central role of the president in implementing diversity agendas, presidents alone cannot shoulder the burden of cultural change; as Birnbaum (1992, 151) observed 'presidents are not the only source of leadership'. As it pertains to implementing the diversity agenda, Kezar, Bertram Gallant, and Lester (2011, 147) identified faculty and staff grassroots leadership to be a vital strategy 'that is aligned with academic culture and institutional methods', and '[allows for] grassroots leaders to operate under the radar'. Aside from Williams (2013) and Williams and Wade-Golden's (2013) work, few scholars have substantively addressed how institutional leaders might best align the efforts of faculty, staff, and students to advance the diversity agenda.

Future research should also address the role of identity in leading the implementation of the diversity agenda. Though little of the literature considered the role of race and gender, Owen (2009) observed that '[diversity] leaders in higher education are not expected to be White males', and that 'men of color are as prevalent as women in [diversity leadership positions]' (186). Recognizing the role both race and gender play, not only as it pertains to the opportunities for the appointment of women and people of color to senior leadership roles (Jackson 2003, 2004, 2008; Jackson and O'Callaghan 2009, 2011), but also the role gender (Eagly and Johannesen-Schmidt 2001; Eagly and Johnson 1990; Young 2004), race (Ospina and Foldy 2009), and the ways in which race and gender intersect (Christman and McClellan 2008; Eagly and Chin 2010; Gasman, Abiola, and Travers 2015; Jean-Marie, Williams, and Sherman 2009; Wolfe and Dilworth 2015) in informing leadership practice is an important consideration for researchers and practitioners alike. As Kezar and Eckel (2008, 396–397) noted, race plays an important role in leaders of color's choice of leadership style, with '[o]ver half of the presidents of color [in their study]', having been weary of transformational approaches, 'because of the way that white stakeholders might perceive [their] choice [of pursuing the diversity agenda] as a personal agenda or self-interest rather than an institutional imperative'.

Institutions of higher learning will continue to face significant challenges in seeking to align policies and practices with their stated commitments to diversity. The work of transforming staid academic cultures and decentralized academic institutions remains a persistent hurdle for leaders to clear, regardless of their institutional position. Although researchers and practitioners alike have long-grappled with the challenge of transforming college and university cultures to become more inclusive and equitable, leading coordinated and intentional initiatives aimed at altering change-resistant institutions remains a heavy burden. Paying close attention to the role and the type of leadership necessary for undertaking the large task of implementing an equity-driven agenda, however, will make the task less burdensome.

Disclosure statement

No potential conflict of interest was reported by the authors.

References

Aguirre, A., and R. O. Martinez. 2002. "Leadership Practices and Diversity in Higher Education: Transitional and Transformational Frameworks." *Journal of Leadership & Organizational Studies* 8 (3): 53–62.

Aguirre, A., and R. O. Martinez 2006. Diversity Leadership in Higher Education. *ASHE-ERIC Higher Education Report* 32 (3). San Francisco, CA: Jossey-Bass.

Anderson, J. A. 2008. *Driving Change through Diversity and Globalization: Transformative Leadership in the Academy.* Sterling: Stylus Publishing.

Antonakis, J., and R. J. House. 2002. "The Full-range Leadership Theory: The Way Forward." In *Transformational and Charismatic Leadership, Volume 2: The Road Ahead,* edited by B. J. Avolio and F. J. Yammarino, 3–33. Bingley: Emerald.

Bass, B. M. 1990. "From Transactional to Transformational Leadership: Learning to Share the Vision." *Organizational Dynamics* 18 (3): 19–31.

Bass, B. M., and R. E. Riggio. 2006. *Transformational Leadership.* 2nd ed. Mahwah, NJ: Lawrence Erlbaum Associates.

Bensimon, E. M. 1993. "New Presidents' Initial Actions: Transactional and Transformational Leadership." *Journal for Higher Education Management* 8 (2): 5–17.

Bensimon, E. M., A. Neumann, and R. Birnbaum 1989. Making Sense of Administrative Leadership: The 'L' Word in Higher Education. *ASHE-ERIC Higher Education Report* 18 (1). Washington, DC: School of Education, George Washington University.

Birnbaum, R. 1988. *How Colleges Work: The Cybernetics of Academic Organization and Leadership.* San Francisco, CA: Jossey-Bass.

Birnbaum, R. 1992. *How Academic Leadership Works: Understanding Success and Failure in the College Presidency.* San Francisco, CA: Jossey-Bass.

Bolman, L. G., and T. E. Deal. 2008. *Reframing Organizations: Artistry, Choice and Leadership.* 4th ed. San Francisco, CA: Jossey-Bass.

Boyatzis, R. E. 1998. *Transforming Qualitative Information: Thematic Analysis and Code Development.* Thousand Oaks, CA: Sage.

Brown, F. W., and D. Moshavi. 2002. "Herding Academic Cats: Faculty Reactions to Transformational and Contingent Reward Leadership by Department Chairs." *Journal of Leadership & Organizational Studies* 8 (3): 79–93.

Burns, J. M. 1978. *Leadership.* New York, NY: HarperCollins.

Christman, D., and R. McClellan. 2008. "'Living on Barbed Wire': Resilient Women Administrators in Educational Leadership Programs." *Educational Administration Quarterly* 44 (1): 3–29.

Chun, E. B., and A. Evans. 2009. Bridging the Diversity Divide: Globalization and Reciprocal Empowerment in Higher Education. *ASHE Higher Education Report* 35 (1). San Francisco, CA: Jossey-Bass.

Cooper, H. M. 1982. "Scientific Guidelines for Conducting Integrative Research Reviews." *Review of Educational Research* 52 (2): 291–302.

Eagly, A. H., and J. L. Chin. 2010. "Diversity and Leadership in a Changing World." *American Psychologist* 65 (3): 216–224.

Eagly, A. H., and M. C. Johannesen-Schmidt. 2001. "The Leadership Styles of Women and Men." *Journal of Social Issues* 57 (4): 781–797.

Eagly, A. H., and B. T. Johnson. 1990. "Gender and Leadership Style: A Meta-analysis." *Psychological Bulletin* 108 (2): 233–256.

Eckel, P. D., and A. J. Kezar. 2003. *Taking the Reins: Institutional Transformation in Higher Education.* Westport, CT: Greenwood Publishing.

Gasman, M., U. Abiola, and C. Travers. 2015. "Diversity and Senior Leadership at Elite Institutions of Higher Education." *Journal of Diversity in Higher Education* 8 (1): 1–14.

Iverson, S. V. 2007. "Camouflaging Power and Privilege: A Critical Race Analysis of University Diversity Policies." *Educational Administration Quarterly* 43 (5): 586–611.

Iverson, S. V. 2008. "Capitalizing on Change: The Discursive Framing of Diversity in U.S. Land-Grant Universities." *Equity & Excellence in Education* 41 (2): 182–199.

Jackson, J. F. L. 2003. "Toward Administrative Diversity: An Analysis of the African-American Male Educational Pipeline." *The Journal of Men's Studies* 12 (1): 43–60.

Jackson, J. F. L. 2004. "Introduction: Engaging, Retaining, and Advancing African Americans in Executive-Level Positions: A Descriptive and Trend Analysis of Academic Administrators in Higher and Postsecondary Education." *The Journal of Negro Education* 73 (1): 4–20.

Jackson, J. F. L. 2008. "Race Segregation across the Academic Workforce: Exploring Factors that may Contribute to the Disparate Representation of African American Men." *American Behavioral Scientist* 51 (7): 1004–1029.

Jackson, J. F. L., and E. M. O'Callaghan. 2009. "What Do We Know about Glass Ceiling Effects? A Taxonomy and Critical Review to Inform Higher Education Research." *Research in Higher Education* 50 (5): 460–482.

Jackson, J. F. L., and E. M. O'Callaghan. 2011. "Understanding Employment Disparities Using Glass Ceiling Effects Criteria: An Examination of Race/Ethnicity and Senior-Level Position Attainment across the Academic Workforce." *Journal of the Professoriate* 5 (2): 67–99.

Jean-Marie, G., V. A. Williams, and S. L. Sherman. 2009. "Black Women's Leadership Experiences: Examining the Intersectionality of Race and Gender." *Advances in Developing Human Resources* 11 (5): 562–581.

Kezar, A. J. 2001. Understanding and Facilitating Organizational Change in the 21st Century: Recent Research and Conceptualizations. *ASHE-ERIC Higher Education Report* 28 (4). San Francisco, CA: Jossey-Bass.

Kezar, A. J. 2005. "What Campuses Need to Know about Organizational Learning and the Learning Organization." *New Directions for Higher Education* 2005 (131): 7–22.

Kezar, A. J. 2007. "Tools for a Time and Place: Phased Leadership Strategies to Institutionalize a Diversity Agenda." *The Review of Higher Education* 30 (4): 413–439.

Kezar, A. J. 2008. "Understanding Leadership Strategies for Addressing the Politics of Diversity." *The Journal of Higher Education* 79 (4): 406–441.

Kezar, A. J. 2014. *How Colleges Change: Understanding, Leading, and Enacting Change.* New York: Routledge.

Kezar, A. J., T. Bertram Gallant, and J. Lester. 2011. "Everyday People Making a Difference on College Campuses: The Tempered Grassroots Leadership Tactics of Faculty and Staff." *Studies in Higher Education* 36 (2): 129–151.

Kezar, A. J., R. Carducci, and M. Contreras-McGavin. 2006. Rethinking the 'L' Word in Higher Education: The Revolution of Research on Leadership. *ASHE Higher Education Report* 31 (6). San Francisco, CA: Jossey-Bass.

Kezar, A. J., and P. D. Eckel. 2008. "Advancing Diversity Agendas on Campus: Examining Transactional and Transformational Presidential Leadership Styles." *International Journal of Leadership in Education* 11 (4): 379–405.

Kezar, A. J., P. D. Eckel, M. Contreras-McGavin, and S. J. Quaye. 2008. "Creating a Web of Support: An Important Leadership Strategy for Advancing Campus Diversity." *Higher Education* 55 (1): 69–92.

Kezar, A. J., W. J. Glenn, J. Lester, and J. Nakamoto. 2008. "Examining Organizational Contextual Features that Affect Implementation of Equity Initiatives." *The Journal of Higher Education* 79 (2): 125–159.

Levy, A., and U. Merry. 1986. *Organizational Transformation: Approaches, Strategies, Theories.* Westport, CT: Praeger.

Ospina, S., and E. Foldy. 2009. "A Critical Review of Race and Ethnicity in the Leadership Literature: Surfacing Context, Power and the Collective Dimensions of Leadership." *The Leadership Quarterly* 20 (6): 876–896.

Owen, D. S. 2009. "Privileged Social Identities and Diversity Leadership in Higher Education." *The Review of Higher Education* 32 (2): 185–207.

Selznick, P. 1948. "Foundations of the Theory of Organization." *American Sociological Review* 13 (1): 25–35.

Tierney, W. G. 1989. "Advancing Democracy: A Critical Interpretation of Leadership." *Peabody Journal of Education* 66 (3): 157–175.

Weick, K. E. 1976. "Educational Organizations as Loosely Coupled Systems." *Administrative Science Quarterly* 21 (1): 1–19.

Williams, D. A. 2013. *Strategic Diversity Leadership: Activating Change and Transformation in Higher Education*. Sterling: Stylus Publishing.

Williams, D. A., J. B. Berger, and S. A. McClendon. 2005. *Toward a Model of Inclusive Excellence and Change in Postsecondary Institutions* [Online]. Washington, DC: Association American Colleges and Universities. Accessed 28 April 2013. http://citeseerx.ist.psu.edu/viewdoc/download?doi=10.1.1.129.2597&rep=rep1&type=pdf

Williams, D. A., and K. C. Wade-Golden. 2013. *The Chief Diversity Officer: Strategy, Structure, and Change Management*. Sterling: Stylus Publishing.

Wolfe, B. L., and P. P. Dilworth. 2015. "Transitioning Normalcy: Organizational Culture, African American Administrators, and Diversity Leadership in Higher Education." *Review of Educational Research* 85 (4): 667–697.

Young, P. 2004. "Leadership and Gender in Higher Education: A Case Study." *Journal of Further and Higher Education* 28 (1): 95–106.

Building the Anti-racist University, action and new agendas

Ian Law

ABSTRACT

This article reviews two decades of work carried out at the Centre for Ethnicity and Racism Studies, University of Leeds in the area of racism and higher education. It introduces key issues and themes in this field and also identifies a seven-point agenda for action. This article provides an overview and agenda-setting account of the theoretical and policy innovations developed by this research team, which provide a contextual background for this volume as a whole. Historical recognition of the role of universities as key sites for the production of racialised knowledge across a range of intellectual fields is an essential starting point. We urge promotion of fundamental de-racialisation and de-colonisation of the academy. This cannot be achieved by self-regulation by the sector or by the setting of minimum legal requirements, it requires strong political, institutional and intellectual leadership, alliance-building and mobilisation.

Introduction: the CERS record, 20 years of research and action

There is an urgent need to interrogate and challenge the historical and contemporary processes of racism, whiteness and Eurocentrism that operate in universities around the world, and particularly in the UK. For nearly two decades, Centre for Ethnicity and Racism Studies (CERS) has been developing a foundational critique of Higher Education Institutions (HEIs) based on a programme of work entitled *Building the Anti-racist University*. CERS was established in 1998, building on the success of Race and Public Policy Unit (RAPP), led by Malcolm Harrison and myself and established in 1992. RAPP focused on issues of racism and ethnicity across a variety of social policy fields in the UK with a particular focus on housing, social security and community and psychiatric care. Bringing together researchers active in this field across many departments at Leeds there was a common set of concerns about racism in the university sector, and at our first meeting a primary collective objective was agreed to work towards building an anti-racist agenda across these institutions. CERS was established as a horizontal, flat, fluid network which facilitated the promotion of research in this field and which did not become an administrative and bureaucratic straitjacket for those involved. Underlying the establishment of CERS was also a collective will to keep the spotlight on racism as a primary field of research, symbolised in the Centre's name. This was significant in its work for anti-racist transformation in HE particularly as a dominant trend in allied research centres in the UK was to jettison the specification of racism as a primary object of critical inquiry in favour of other foci including ethnic relations, migration and identities. CERS positions itself firmly within the long sociological tradition placing race and racism at the centre of the making of Western modernity, from Du Bois, Cooper, Cesaire and Fanon to contemporary theorists including Hall, Hesse, Collins, Goldberg, Glissant and Winant.

This article provides an overview of this work and makes the case for the global transformation of HEIs towards this goal. The wider CERS goal of an *Anti-racist Future* was recently set out in our Manifesto (see preface to Sian et al. 2013).

Building an anti-racist future in HEIs

A vision of the future is in sight – the total transformation and dismantling of racism – through the mobilisation of a series of global transformations in the way the world works. Yet, we are beset on all sides as racism 'surges around us'. Regimes across the world live in a perpetual state of denial. Racism is not here these states cry, from China to the Russian Federation, from Myanmar to Mexico and from Hungary to Lebanon, racism is over there, somewhere else, or just over. Despite the advances that have been made and the dangers of overstating historical optimism, for many, racism is incomprehensible. There is a chronic crisis in grasping how this social force works in the world today.

Many academics, university administrators and Vice Chancellors also fail to grasp the significance and power of racism in their own organisations and practices and lack the motivation and creativity necessary to respond to this challenge.

Despite the introduction of race relations legislation in 2000 which required UK HEIs to produce race equality documents and which embody a potentially far-reaching set of requirements, it may be argued that they fall woefully short of an agenda that could emerge from a more fundamental and serious consideration of a combination of anti-racist, multicultural and racial equality questions and issues. The privileging of race equality as diversity for institutional policy-making as a result of legal duties also carries with it a downplaying of alternative policy priorities. Promoting multiculturalism or anti-racism as a policy goal may involve very different institutional questions and appropriate strategies. Historically, universities have largely catered for white privileged males, and a white, elitist, masculinist, heterosexist, able-bodied and Eurocentric culture still pervades many of the older established institutions as well as 'new universities' today.

Although there has been considerable research into race equality issues in schools in the UK, there has been less analysis of 'race equality' and racism in HEIs. This is perhaps indicative of the complacency that has pervaded the higher education sector. There is nevertheless a series of emerging concerns. These relate to ethnic inequalities in student access, racial discrimination by admissions tutors, the racist experiences of Black and Asian students on entering HEIs, disillusionment with the lack of diversity in the teaching and learning environment, racist discrimination in marking and assessment, racism in work placements and race discrimination in graduate access to employment. In addition, racism and racial discrimination suffered by staff in universities are increasingly being exposed in individual cases and organisational audits. Evidence from academics and support staff in the old universities revealed that racialised tensions are common in universities, with Black and minority ethnic staff often experiencing racial harassment, feeling unfairly treated in job applications, and believing institutional racism exists in the academic workplace. The development of subject areas and disciplines has also been critiqued as reproducing and reinforcing a Eurocentric world view which peripheralises and fails to value that which is seen to lie 'outside' the West. Relevant questions to ask in this respect are: are the literatures, music, arts, histories and religions etc. of 'non-Western'/'not-white' peoples periphalised and tokenised in the curriculum? Are the literatures, music, arts, histories and religions etc. of 'non-Western'/not-white peoples positioned as inferior, primitive? And are cultures etc. other than the dominant culture of the HEI valued, displayed, celebrated, promoted? Staff and departments should be mindful to consider the inclusion and integration of voices, perspectives, works and ideas that come from beyond a 'white', 'Eurocentric' core.

There are a number of issues to be mindful of in terms of considering the learning environment and the needs of students. The process of learning needs to be inclusive and should consider the needs of all learners in terms of ethnicity, gender, disability, religion and so on. Lecturers, tutors etc. should be aware that their own expectations of students may be based on stereotypes and assumptions about what particular Black and minority ethnic groups 'are like' or the kinds of expected aptitude for particular activities, subjects, approaches etc. As such, care should be taken to avoid making assumptions and having expectations about students based on these stereotypes. International students are particularly vulnerable here as assumptions of academic inferiority often circulate with reference to students from non-Western countries.

It is time for HEIs in the UK to re-conceptualise their role and responsibilities in a contemporary multicultural society. Experience has shown that race equality will not be achieved easily and it is unlikely to be attained through the implementation of an all-encompassing 'equal opportunities' programme. This has led to the marginalisation of race equality initiatives after the initial 'kick-start' that the legislation gave has faded. There is a need to create an anti-racist culture within HEIs in general, and, most urgently, in the older established institutions in order to challenge entrenched systems of white privilege. Progress will only occur if anti-racism becomes part of the professionalism of staff, as well as the code of conduct for students and is embedded in working relationships with the external community. Success is dependent on the support and goodwill of staff at all levels. Many staff and students in universities have ambivalent or hostile attitudes to anti-racist and race equality strategies, as they believe that the system is 'already fair' and that any new measures will favour minority ethnic groups over white people. Institutional cultures are, however, rapidly changing and the value of the changing legal context has undoubtedly been a significant factor in promoting progress in this field.

The CERS stream of work on racism and the university sector has included the following:

1993: qualitative and quantitative study of ethnic monitoring of University admissions at Leeds was carried out identifying racial inequalities in the success rates of undergraduate admissions and widely differing, subjective perceptions in admissions practice (Robinson et al. 1993).

1996: wider review of racial inequalities in university admissions published identifying racial discrimination and the insularity of the HEI sector from anti-racist developments, myths of academic liberalism and associated denial of racism on campus, hostility to prescription and arrogance and complacency in the face of racial and ethnic inequalities (Law 1996).

2002: HEFCE Innovations Fund project the *Institutional Racism Toolkit* launches a web-based resource for UK HEIs (Turney, Law, and Phillips 2002). The introduction of the Race Relations (Amendment) Act 2000 for the first time placed a statutory duty on HEIs in the UK to eliminate racial discrimination and promote racial equality. In many institutions there was a knowledge vacuum and little guidance on how to move forward. This research project was designed to fill this gap. The research, carried out between 2000 and 2002, was co-authored by Ian Law (University of Leeds (Univ. Of Leeds 1991-), Deborah Phillips (Univ. of Leeds, 1988–2008), now University of Oxford) and Laura Turney (Univ. of Leeds, 2000–2002, now Scottish Government), and supported by the HEFCE Innovations Fund.

The project conducted a review of organisational dimensions of institutional racism and race equality in the HE Sector using the University of Leeds as its case study. Email surveys of 2000 staff and 2500 students and 30 interviews with heads of schools and administrative units were carried out together with analyses of ethnic origin data-sets on admissions and employment coupled with documentary analysis of policy and practice. One senior academic commented 'the University hierarchy is very white, male, suited and middle-aged, in both composition and culture' and further that 'racism is not overt but subtle in its manifestations – assumptions made and language used in documentation and professional dialogue'. Interview and survey data from the Leeds study certainly indicate that large numbers of key staff are opposed to understanding an HEI as an institution in which race discrimination is embedded across policy, practice and organisational culture. Findings confirmed the prevalence of racist discourses and incidents in HEI settings with approximately 25% of staff and students identifying these practices. Major spheres, where no attention had been given to these issues included, for example, teaching and learning, and contracting and purchasing, demonstrating the need for fundamental organisational change. The toolkit combines research evidence and new instruments for organisational analysis. The research included the development of a new theoretical framework synthesising racism, whiteness and Eurocentrism which was used to interrogate HEI policy and practice. The toolkit applies these concepts to the main organisational areas of HEI activity including leadership and management, teaching and learning, employment, research, contracting and external relations. The toolkit also provides a set of methodological and management tools for investigating, understanding and intervening in institutional racism in HEIs. This resource was launched online in 2002 at a major national conference in Leeds which for the first time addressed racism in the HEI sector bringing together 140 practitioners, academics, researchers, community and trade union representatives and policy-makers. It was praised as 'a most valuable and innovatory resource for the higher education sector' (Joyce Hill, former Director, Equality Challenge Unit). The aim of this event was to turn the lack of focus on this issue into a policy problem, and to propose solutions which went significantly beyond the meeting of legal minimum requirements, and also to begin the process of long-term dialogue with HEI's to achieve institutional change. It has also made a significant impact on this field of study, following a programme of dissemination and user engagement, with wide recent citation.[1]

The specific interventions identified here include the development of a toolkit of resources to enhance professional practice and the stimulating of new debate about institutional racism and the output of appropriate strategies by a large number of HEIs who have used the toolkit as a foundation. It also shows accumulating impact on other sectors of public policy through stimulating debate about the renewal of anti-racist strategies via the Council of Europe and the European Commission.

Furthermore, interest from practitioners in the public and voluntary sectors in the UK and utilisation of our output in developing organisational strategies indicates impact beyond the discipline. Overall, this work has been described as 'ground-breaking' by Mirza (Open University 2004). This work has also achieved recognition through invitations to contribute to policy development of racial equality and anti-racist strategies through the production of innovative guides and web-based resources, for example for the UK Teaching Quality Enhancement Fund team.

The value of the toolkit in the development of racial equality strategies in higher education outside the UK has been confirmed by a variety of institutions, for example in South Africa by Nelson Mandela University. In the USA, the American Sociological Association's Minority Opportunities through School Transformation programme confirmed the value of the HEI toolkit in promoting debate and developing interventions to reduce racial inequalities in access to higher education.

2004: *Institutional Racism in Higher Education* edited book published reporting leading edge research on racism in HEIs (Law, Phillips, and Turney 2004) including a comparison of the similarities in policy failure by a police service and an HEI in the Midlands, an assessment of Gypsy and Traveller access to HEIs and proposals for a transformative curriculum.

2007: study on *South Asian women in Higher Education* funded by the Joseph Rowntree Foundation published (Bagguley and Hussain 2007). This study compared the aspirations and experiences of British Indian, Pakistani and Bangladeshi origin women. It explored how they balance their education with plans for marriage, and their experiences of racism and Islamophobia at university and elsewhere. It analysed the barriers to higher education arising from institutional, financial and community factors and the ethnic segregation that appears to be emerging among the traditional old universities and the new universities.

2008: invited contribution made by CERS to the Council of Europe *Intercultural Dialogue on Campus* initiative (Law 2009), this included assessment of the causes of intercultural conflict on campus and their effects, and also assessment of the value of the anti-racist toolkit to European debates in this field. The Council of Europe recognised the value of the *Building the Anti-racist University* toolkit as evidenced by an international invitation to develop the significance of these research messages for the Council of Europe in 2008, in the context of the European Year of Intercultural Education, and present these at a conference in Strasbourg. This resulted in a subsequent keynote presentation and a Council of Europe publication which highlighted the causes of intercultural conflict and how new strategies to address these could be implemented on campuses across Europe. This research has stimulated new debate in this field and influenced Council of Europe policy and practice approaches to intercultural dialogue on campus and indicates the increasing international recognition of this work.

2009: *WUN (World Universities Network) White Spaces* was established by Shona Hunter with a key focus on interrogating whiteness in academia. This network includes academics, postgraduate students and practitioners from across 23 different countries: Argentina, Australia, Brazil, Canada, South Africa, USA, New Zealand, Germany, Sweden, Switzerland, France, Greece, Finland, Italy, Spain, Lithuania, Netherlands, Norway, Denmark, Israel,

Mexico, Portugal, UK and 17 disciplines across the humanities, health, psychology and social as well as some natural sciences.

2012: Colloquium on *Global Research on the Black Male Educational Pipeline: International Perspectives to Inform Local Solutions* held at the University of Leeds, a collaboration between Shirley Anne Tate, CERS and James L. Moore III, Associate Vice Provost, Ohio State University and Jerlando F.L. Jackson, Director of Wisconsin's Equity and Inclusion Laboratory, University of Wisconsin (UW). This showcased interventions to improve Black male performance on campus. The colloquium aimed to share knowledge gleaned from research on Black/of colour boys and young men at different stages of the educational pipeline, to share approaches to community engagement, access, retention at UG level and progress to PGT/PGR status; to work with students on issues of racism on campus; to enable students to build a portfolio of skills and develop a brand for entry to graduate level education and beyond into the labour market; and to empower students through engagement with and mentoring from senior academics both from the UK and the USA. This is intended to be the first of such international colloquia initiated by UW-Madison and has been continued yearly in different national locations.

Examples of innovatory programmes from the UW-Madison from which we can learn globally are based on working with students in school alongside peer mentoring and academic mentoring while at university:

(a) The Pre-College Enrichment Opportunity Programme for Learning Excellence (PEOPLE) programme

Began in 1999 and is based in the Office of the Vice Provost for Diversity and Climate. It is a central plank in the UW-Madison's approach to enabling access. It is a year-round learning experience over 6 years until high school graduation that engages under-represented youth in both middle and high school who are considering college education in subjects right across the University departments ranging from, for example, Performance Studies to STEM. In the in-school intervention this programme is a combination of curriculum enhancements in the summer which build academic skills such as Maths, English, study skills and writing skills development as well as workshops in the biological and physical sciences, engineering, biomedical research, health sciences and law; and for older students an internship/research experience for learning and applying methods of scientific inquiry, analysis and research in the humanities and social sciences; as well as experience and exposure to various professional fields through placements within and outside of the university. One hundred per cent of students in the programme graduate from high school and 95% enrol in higher education. PEOPLE students admitted to the UW-Madison normally also complete the summer bridge-to-college programme. PEOPLE scholars who graduate from UW-Madison are prepared to fill management and technical positions in the public and private sectors, pursue graduate studies leading to careers in academia or other professions and assume leadership positions with civic and community institutions.

(b) The Posse Programme

The Posse Programme exists in several universities across the USA and aims to develop peer mentoring relationships among students either on campus or across campuses within the USA. Peer mentoring runs from the beginning of UG level up to and beyond PGR level.

2013: *Racism, Governance and Public Policy: Beyond Human Rights* (Sian, Law, and Sayyid 2013). The wider application of this work to European public policy has informed the development of an EU FP7 project *The Semantics of Tolerance and (Anti-) Racism in Europe: institutions and civil society in a comparative perspective.* This project extended the reach of the *Building the Anti-racist University* toolkit approach across public policy areas. The research, co-authored by Ian Law, Salman Sayyid (CERS, University of Leeds, 2005-) and Katy Sian (University of Leeds 2010–2012, now University of York), includes analysis of the discursive construction of Muslims in three contexts: the workplace, schooling and the media. It is informed by a fundamental critique of both the 'post-racial' and the limitations of human rights strategies and identifies the ongoing significance of contemporary racism in governance strategies and develops a new radical agenda for addressing these processes.

2013: *Building the Anti-racist University* international conference held at the University of Leeds which brought together multiple international partners, including representatives from HEIs in Brazil, Canada, USA, Europe and South Africa to continue the process of long-term dialogue, agenda setting and the development of policy solutions. A set of papers from this conference constitute this current collection.

2014: Since the implementation of a statutory obligation on implementing racial equality in public sector organisations was introduced and associated sector wide activity to promote action in this area was carried out by the Equality Challenge Unit, trade unions and ourselves over 300 institutions in the UK established racial equality strategies and have improved experiences particularly for black and minority ethnic students (National Students Survey 2002–2012, HEFCE 2012). Stimulating institutional change towards the construction of the Anti-racist University was the aim of the CERS toolkit. This approach was concerned to develop a maximal, transformative approach to institutional change, rather than a minimal meeting of legal obligations. This work has informed the development of Leeds University's Racial Equality Scheme, and many others across the sector within and outside the UK. But, progress in this field has slowed and focus on this goal has dissipated both at Leeds and across the sector.

At our own institution, the University of Leeds' Equality Objectives 2012–2016 highlight persisting racial and ethnic inequalities in the representation of Black and minority ethnic staff at leadership and management levels, and also in differential success rates of student admissions and degree attainment. The University's Single Equality Scheme 2009–2012 sets out the steps taken to achieve minimal legal compliance with the statutory duty on race equality. Some areas of progress include a significant increase in the representation of black and minority ethnic students across the institution, staff training, consultation, data collection, Black History month annual programme of activities and in purchasing. In our view these policy statements are inadequate and do not reflect the necessary institutional effort required to establish the University as a global leader in this field. We propose a review and refocusing of strategy and action in this sphere, an injection of appropriate resources to support innovative action, and the development of a new strategy that is not framed by legal obligations but by intellectual, moral and institutional goals.

In sociology and social policy we have developed an intensive field of research and teaching activity in the field of racism and ethnicity studies. There was a pathway to pursue this field of knowledge across our programmes of study at every level, from Foundation (Year 0) through three years of undergraduate programmes, an MA programme and beyond to doctoral and postdoctoral levels. We have also built up a cohesive team of research and teaching active staff in this field comprising Shirley Anne Tate, Ian Law, Paul Bagguley, Salman Sayyid, Yasmin Hussain, Shona Hunter, Rodanthi Tzanelli, Adrian Favell, Roxanna Barbulescu and Richard Tavernier, a network of researchers across the University and wider international networks through projects including two EU FP7 projects, EDUMIGROM, TOLERACE and White Spaces. Pursuing this stream of work discussed here, in the sector and institutions in which we work, is a core priority for CERS.

There is a new focus at CERS: global racism studies, with new books, a dedicated book series and other outputs, a new Mapping Global Racisms Research Archive of working papers. Here theoretical innovation involves making a theoretical break, incorporating the new conception of polyracism, which involves moving beyond the partial, limited account of global racialisation stemming from the critical race tradition in arguing for the application and extension of this critique across the planet, historically and geographically. Why restrict our deconstruction of racial logics to the operation of Western capitalist modernity? This arbitrary decision has serious consequences in putting many polities and contexts out of critical sight and deeming them as unworthy of interrogation, for example, pre-modern and post-colonial regimes in North Africa and many Communist contexts. The recent exposure of brutality, violence and murder driven by the North Korean state's regulation of racial purity in relation to children of mixed North Korean and Chinese heritage where a prison camp mother was ordered to drown her own baby illustrates this problem (Guardian 2014). The exposure of the North Korean regime's claim to be the 'cleanest', 'purest' race, influenced by Japanese fascism, has only recently received scholarly attention (Myers 2011). Inattention to the proliferation of non-Western racial modernities is also evident in the lack of interrogation of the Soviet Union and the Chinese People's Republic. Contemporary racisms in Morocco, Algeria, Libya and the Lebanon, together with examination of anti-gypsyism in Turkey and the Middle East are some of the national contexts which illustrate the importance of a non-Western focus of study in this field.

This new theory of polyracism proposes a conceptualisation of the historical development of multiple origins of racism in different regions and forms, as opposed to the monoracism arguments positing a linear diffusion of Western racisms from the classical world onwards and outwards. This argument also involves examining racial interconnectivities, crossings and connections, for example, in the development of pre-modern racial discourse in the Mediterranean region, which is deployed here to unsettle, counter and disrupt the parochial insularity of Eurocentric accounts of the historical development of racism. So, rather than racism being the product solely of Western modernity, polyracism theory argues that it is also pre-modern (proto-racism), non-Western, non-capitalist (Communist) and the product of other varieties of modernity. This is not to argue that racism is always and inevitably everywhere. It is the product of, and operates under specific conditions in specific places, cultures and polities. The concept of racial conditions is used to elaborate where and in what ways contemporary racisms operates.

Polyracism theory builds on work elaborated in *Red Racisms* (Law 2012), with particular reference to racial regimes in Russia, Cuba, China and four states in Central and Eastern Europe, and recently elaborated here in relation to selected dimensions and aspects of the Mediterranean region and its histories (Law et al. 2014). Communist regimes are rooted in 'solid' modernity with grand narratives and a rational belief in progress through highly controlled use of technology, bureaucracy and military power and they too have their racialised hierarchies and racialised internal enemies and targets of hate, and are ordered and regulated by identifiable racial logics in state governance. The complacency, arrogance and hypocrisy of these regimes declaring themselves immune to racism has for too long been hidden from scrutiny. Polyracism theory has also been elaborated in the Caribbean context (Tate and Law 2015). The Caribbean is characterised by some of the most complex interactions between previously divergent populations from the extensive Mesoamerican migrations in pre-Columbian times onwards (Moreno-Estrada et al. 2013). The dilemmas and directions of historical and contemporary debates over what work whiteness, blackness and mixedness do in the Caribbean context is a central theme here. Through this Caribbean triad the power of racialisation and its long reach is held up to critical scrutiny. The Caribbean is a complex context and this book cannot do justice to all parts, peoples and places, although it does aim to establish and interrogate some key overarching regional relational racial dynamics and processes together with attention particularly to the island, rather than mainland Caribbean and a set of selected case study contexts including Jamaica, Trinidad and Tobago, Cuba, Puerto Rico, the Dominican Republic and Haiti. Racial Caribbeanisation is the process of ethno-racial domination of this region rooted in European colonialism encompassing the conquest and genocide of the Amerindian peoples, the enslavement and exploitation of Africans, use of indentured labour and the embedding of racial and ethnic hierarchies in post-colonial, post-independence contexts. The interrogation of this process is the central focus of this book. This book has sought to delineate some of the racial trajectories of Caribbean states which include increasing concentrations of white wealth and financial power in small island locations, multiracialised national projects of inclusion, intensifying colonialisms, aspirational whiteness, the pursuit of racial Americanisation and vehement anti-blackness. This proliferation of racial forms and conditions indicates the contemporary power and intensity of the waves of polyracial neoliberalism which perpetually wash across the Caribbean seascape. Polyracism theory is a key building block in the ongoing programme of research based at CERS concerned with the theorisation and specification of global racialisation (Law, 2009, 18) under the broad heading of Mapping Global Racisms. This project is also informed by research-led teaching and the output produced by undergraduate and postgraduate social scientists at the University of Leeds, who have contributed to the Mapping Global Racisms Research Archive (available at cers.leeds.ac.uk). This consists of case study work examining many racial states outside the UK. We have as yet a very limited, partial, uneven account of world racisms and there remains much to document, criticise and challenge in building systematic theory, evidence and multiple anti-racist futures.

Neoliberalism effectively masks racism through its value-laden moral project, camouflaging practices that are anchored in an apparent meritocracy, making possible a utopic vision of society that is non-racialised. The operation of the free reign of markets provides a political and economic terrain, which facilitates the double movement of resignified, rebranded cultures and identities, new segregations, divisions and exclusions. Placing processes of race and racialisation as a 'foundational pillar' (Goldberg 2008) of modernising globalisation enables them to be identified as constituting a new and renewing pattern of modern state and regional arrangements for managing populations. The increasing shift to neoliberal states, where their role becomes one of securing conditions for the maximisation of privatised interests and corporate profits, has provided a new terrain for configurations of race. The renewed critical debate about the role that neoliberalism plays in contemporary forms of racialisation provides an important dimension in developing analysis of policy and governmentality (Bhattacharya 2013; Goldberg 2008; Gopalkrishnan 2007; Hall 2011). Neoliberalism has provided a hegemonic framework within which people have been bound into political projects which carry through a range of strategies and techniques of governance and managerialism. These involve securitisation, military occupation, penalising the poor and creating 'infeartainment' as fear is mobilised by states – a key emotional political strategy. The transformation to forms of neoliberal governmentality has had profound consequences for universities and racialised groups. Here the work that such discourse does is to bury racialised forms of hierarchical social relations, reinterpreting these, for example, as individualised 'inadequately mobilised social capital', which exposes these populations to new forms of exploitation and containment, and market-driven differentials in assessments of human value and human need. Any challenge to these arrangements must therefore engage with the political projects of polyracial neoliberalism, remaking states and institutions anew in pursuit of deracialisation, just as this new form of governmentality seeks to transform prior types of state and institutional configuration.

A seven-point agenda for change

Leadership and restoring antiracism as a foundational intellectual project

Historical recognition of the role of universities as key sites for the production of racialised knowledge across a range of intellectual fields is an essential starting point. As Biller (2009) confirms the marking out of the peoples of the world between the polarities of blackness and whiteness was 'drummed into the minds of university graduates of the Middle Ages' and beyond, for example into the laboratory practices of genomics research (Tutton 2007). We urge promotion of fundamental de-racialisation and de-colonisation of the academy. This cannot be achieved by self-regulation by the sector or by the setting of minimum legal requirements, it requires strong political, institutional and intellectual leadership. Political intersectionality has been key to the formation and success of abolition, anti-colonialism, anti-apartheid, civil rights and many other anti-racist movements and will be key in this wider project also.

Widen the debate

The debate over race and higher education in the UK is too narrow being focused on the important issues of undergraduate access and academic employment. Until this debate is widened to address the core business of research and teaching, impact will remain marginal.

Promote the Building the Anti-racist University good practice model

We have provided a good practice model for organisational change built on key principles of challenging racism, whiteness and Eurocentrism across all areas of university activity, which takes the debate way beyond the meeting of minimum legal requirements and we vigorously advocate its implementation, keeping a strong single strand focus on antiracism and racial justice.

Arrest the marginalisation of these debates

Institutional attention to issues of racial justice is being lost in the university sector, and elsewhere with the move to wider equality, human rights and widening participation agendas. A new debate engaging with issues of affirmative action, racial justice/equality targets and the transformation of racialised institutions is needed.

Cross-sectoral learning

The university sector has been one of the last to address issues of racism and ethnicity, due to the reasons stated above. We need to recognise how we arrived at the present. Therefore, it is important for this sector to learn from other sectors in terms of how to implement fundamental change and to avoid the mistakes made elsewhere.

Cross national learning

In the context of our international networks and knowledge transfer detailed above, there are many lessons to be learned from elsewhere. For example, new developments in affirmative action in Brazil, in educational interventions to improve Black male performance on US campuses and interventions in challenging whiteness in South African HEIs can provide useful lessons for the UK. We advocate the creation of an international network concerned with *Building the Anti-racist University.*

New resourced initiatives are desperately needed

Changing the mainstream will be slow, we advocate resourcing of new appointments, programmes of study, research networks and learning and teaching initiatives concerned with addressing the issues raised in this article. New initiatives are urgently required to lead the way forward.

Note

1. For example the Toolkit is included in: (a) St. Andrew's University *Racial Equality and the Curriculum Staff Guide* (2013) (http://www.st-andrews.ac.uk/staff/policy/tlac/equalitydiversity/racialequality/). (b) Birkbeck, University of London, *Criminology and Criminal Justice Staff Guide* (2012). (c) Plymouth University's *7 Steps to Adopting Culturally Inclusive Teaching Practices* (2010), Newcastle University's School of Medical Sciences *Education Development Resources* (2013) (http://www.medev.ac.uk/resources/506/project/). (d) Institute for Education, University of London *Respecting Difference, good practice guide for PGCE Tutors in issues of race, faith and culture* (2008) (http://www.ioe.ac.uk/RespectingDifference.pdf). (e) University of Huddersfield's Race Equality Resources. 2013. (http://www.hud.ac.uk/equality/race/).

Disclosure statement

No potential conflict of interest was reported by the author.

References

Bagguley, P., and Y. Hussain. 2007. *Moving on up, South Asian Women and Higher Education*. Stoke on Trent: Trentham.

Bhattacharya, Gargi. 2013. "Racial Neoliberal Britian." In *The State of Race*, edited by N. Kappor, V. Kalra, and J. Rhodes. Basingstoke: Palgrave.

Biller, Peter. 2009. "Proto-racial Thought in Medieval Science." In *The Origins of Racism in the West*, edited by Miriam Eliav-Feldon, Benjamin Isaac, and Joseph Ziegler. Cambridge: Cambridge University Press.

Goldberg, D. T. 2008. *The Threat of Race, Reflections on Racial Neoliberalism*. Oxford: Blackwell.

Gopalkrishnan, N. 2007. "Neo-liberalism and Infeartainment: What does a State do?" In *Racisms in the New World Order: Realities of Culture, Colour and Identity*, edited by H. Babacan and N. Gopalkrishnan. Newcastle: Cambridge Scholars.

Guardian. 2014. *North Korea Human Rights Abuses Resemble Those of the Nazis*, February 18.

Hall, Stuart. 2011. "The Neo-liberal Revolution." *Cultural Studies* 25 (6): 705–728.

Law, I. 1996. *Racism, Ethnicity and Social Policy*. Hemel Hempstead: Harvester/Wheatsheaf.

Law, I. 2009. "Defining the Sources of Intercultural Conflict and their Effects, in Bergan, Sjur and Restouiex." In *Intercultural Dialogue on Campus, Council of Europe Higher Education Series No. 11*, edited by Sjur Bergan and Jean-Phillipe Restouiex. Strasbourg: Council of Europe. http://book.coe.int/EN/popupprint.php?PAGEID=36&produit_aliasid=2415.

Law, I. 2012. *Red Racisms, Racism in Communist and Post-communist Contexts*. London: Palgrave.

Law, I. with A. Jacobs, N. Kaj, S. Pagano, and B. Sojka-Koirala. 2014. *Mediterranean Racisms, Connections and Complexities in the Mediterranean Region*. London: Palgrave.

Law, I., D. Phillips, and L. Turney, eds. 2004. *Institutional Racism in Higher Education*. Stoke-on-Trent: Trentham Press. http://trentham-books.co.uk/acatalog/Trentham_Books_Institutional_Racism_in_HigherEducation_277.html.

Moreno-Estrada, A., S. Gravel, F. Zakharia, J. L. McCauley, J. K. Byrnes, C. R. Gignoux, P. A. Ortiz-Tello et al. 2013. "Reconstructing the Population Genetic History of the Caribbean." *PLoS Genetics* 9 (11).

Myers, B. R. 2011. *The Cleanest Race, How North Koreans see Themselves and Why it Matters*. Brooklyn, NJ: Melville House.

Open University. 2004. *Innovations*. Milton Keynes: Open University.

Robinson, P., M. Harrison, I. Law, and J. Gardiner. 1993. *Ethnic Monitoring of University Admissions: Some Leeds Findings*. Social Policy and Sociology Research Working Paper 7. Leeds: University of Leeds.

Sian, Katy, Ian Law, and S. Sayyid. 2013. *Racism, Governance and Public Policy, beyond Human Rights*. London: Routledge.

Tate, S., and I. Law. 2015. *Caribbean Racisms, Connections and Complexities in the Caribbean Region*. London: Palgrave.

Turney, L., I. Law, and D. Phillips. 2002. *Institutional Racism in Higher Education, Building the Anti-racist University: A Toolkit*. http://www.sociology.leeds.ac.uk/assets/files/research/cers/the-anti-racism-toolkit.pdf.

Tutton, R. 2007. "Opening the White Box: Exploring the Study of Whiteness in Contemporary Genetics Research." *Ethnic and Racial Studies* 30 (4): 557–569.

Addressing dualisms in student perceptions of a historically white and black university in South Africa

Ronelle Carolissen and Vivienne Bozalek

ABSTRACT

Normative discourses about higher education institutions may perpetuate stereotypes about institutions. Few studies explore student perceptions of universities and how transformative pedagogical interventions in university classrooms may address institutional stereotypes. Using Plumwood's notion of dualism, this qualitative study analyses unchallenged stereotypes about students' own and another university during an inter-institutional collaborative research and teaching and learning project. The project was conducted over 3 years and 282 psychology, social work and occupational therapy students from a historically black and white institution in South Africa, participated in the study. Both black and white students from differently placed higher education institutions display prejudices and stereotypes of their own and other institutions, pointing to the internalisation and pervasiveness of constructions and hegemonic discourses such as whiteness and classism. It is important to engage with subjugated student knowledges, in the context of transformative pedagogical practices, to disrupt dominant views and cultivate processes of inclusion in higher education.

Introduction

The question of how to implement anti-racist interventions in higher education remains a pertinent global issue. In South Africa, national documents such as the Ministerial Committee on Transformation and Social Cohesion (Department of Education 2008) and the Report on the Summit on Higher Education (2010) foreground the importance of diversity and inclusion in the higher education sector with reference to race, gender and class. Transformation charters of universities such as Stellenbosch University (SU), and the University of the Western Cape (UWC) are additional and specific institutional documents that refer to visions for diversity on specific South African campuses. In spite of this component being enshrined in these national (Department of Higher Education and Training 2010) and institutional policy frameworks (SU and UWC), little literature exists about engaging in dialogue about difference from a student perspective in higher education (HE) institutions (Cross and Johnson 2008; Denson and Chang 2009; Jansen 2009; McKinney

2004, 2005, 2007), although some literature exists on this topic in school contexts (Hemson 2006; Soudien 2012; Stoughton and Sivertson 2005). The literature that does exist on engagement with difference in HE contexts, typically focuses on particular institutions (Cross and Johnson 2008; Robus and MacLeod 2006; Steyn and van Zyl 2001) with few studies focusing on students engaging in dialogue about difference *across* universities, especially in pedagogical contexts (Leibowitz et al. 2012). This is significant as social constructions (Burr 1995; Gergen and Gergen 2003) and hegemonic discourses of higher education institutions have implications for the perpetuation of stereotypes about institutions and consequently, the relative value that students place upon their own and other institutions. Existing literature furthermore draws on a range of theoretical perspectives; discursive and critical race theory (Robus and MacLeod 2006; Steyn and van Zyl 2001), conceptual frames of campus membership that draw on Nancy Fraser's cultural recognition (1997), agency and Bourdieu's habitus (1986). Few theoretical orientations focus specifically on dualisms and dynamics inherent in dualisms that appear central to normative thinking about difference.

As a group of South African higher educators, we were concerned about the history of minimal inter-professional and inter-institutional engagement between students from psychology, social work and occupational therapy (human service professions), particularly across historically advantaged/white institutions such as Stellenbosch University (SU) and historically disadvantaged/black universities such as the University of the Western Cape (UWC). Students from these two universities rarely have opportunities to engage with each other. We designed a collaborative teaching and learning project to develop an inter-institutional engagement space. Before students commenced the course, only 25% of them had visited each others' universities in spite of the universities being geographically close. We perceived the lack of engagement across institutions as having negative consequences for teaching and learning, as students and educators have limited opportunities to experience and explore difference in relation to themselves, others, their curricula, disciplines and institutions. In the absence of such plurality of perspectives, inter-institutional and interdisciplinary dualisms remain unchallenged.

In this article, we examine how the notion of dualism may be relevant for thinking about issues of difference in higher education, with specific reference to UWC and Stellenbosch student perceptions about the Universities of the Western Cape and Stellenbosch. We contend that the characteristics of dualism outlined by the feminist philosopher Val Plumwood (1993, 2002) may be helpful in attempting to challenge or address dualisms and issues of subjugated constructions of universities among students *in* pedagogical contexts.

Current literature in international and South African contexts that explore student perceptions of universities focus on experiences of teaching and learning (Bartram and Bailey 2009; Ruohoniemi and Lindblom-Ylänne 2008), especially during the first year (Brinkworth et al. 2009; James, Krause, and Jennings 2010; Leibowitz, van der Merwe, and Van Schalkwyk 2009). There is also a body of literature which may be considered to fall into the neoliberal discourse of 'climate surveys' (Brown and Mazzarol 2009) that assesses popularity of universities and the potential that universities have to attract students as paying clients and consumers (Ancis, Sedlacek and Mohr 2000; Brown and Mazzarol 2009). Very little literature engages with students' understanding and engagement in dialogue about difference, especially as it relates to students' perceptions of universities (Cross and Johnson 2008). Student perceptions are important to consider as they are likely to reflect normative assumptions about universities, which may affect student decisions about further study, their self-conceptions and their anticipated career trajectories. Both structural contexts and associated discourses about higher education provide contexts for stereotyping and associated dualisms to arise and it is therefore important to consider how dualisms may frame student perceptions of their own and others' institutions.

Contexts: differentiation in South African HE

The higher education sector, like other levels of education has been, and continues to be, deeply affected by its apartheid past. Before the democratic dispensation in 1994, the South African higher education system was largely constructed as being divided into either historically advantaged or white institutions (HAIs/HWUs) or historically disadvantaged or black higher education institutions (HDIs/HBUs). Subsequent to this, during the period of 2000–2005, the 36 higher education institutions in South Africa were merged and developed into a stratified and differentiated higher education system of 23 higher education institutions in 2012 (Cooper 2015). The two institutions which are focused on in this paper are a historically white advantaged institution and a historically black and disadvantaged institution in the Western Cape region of South Africa that were excluded from these mergers. The higher education institutions in South Africa are now classified as 11 universities, 6 comprehensive universities offering a mixture of traditional and vocational programmes and 6 universities of technology (Bozalek and Boughey 2012; Cooper 2015). More recent research has identified three differentiated groups of institutions characterised by patterns of inputs and outputs of research and postgraduate students. It is marked that the top five of these differentiated institutions are previously historically white or advantaged institutions and that the bottom eleven are either merged universities of technology or previous historically disadvantaged institutions, continuing the legacies of research and postgraduate outputs that were initiated in the apartheid era (Bozalek and Boughey 2012). Thus the effects of apartheid in terms of the governance and resources available to differently categorised institutions continue into the current era. In addition, black students who can both afford to study at HAIs and who meet the more stringent academic standards tend to choose to study at HAIs while working class students who don't meet the academic requirements study at HDIs. This has resulted in many institutions remaining predominantly monocultural in terms of race and class categories with little communication between students and staff from these institutions. The description of the higher education context shows how South African higher education institutions are still marked by dualistic thinking (and structures) along racialised and class faultlines.

Essentialist conceptualisations of difference regarding race, class and gender are based on dualisms. Plumwood (1993, 2002) argues that central to the construction of dualism is the idea of two polar opposites, which are hierarchised. One pole is always inferior to the other and the other represents the desirable norm with no possibility of continuity or mutuality between these two sides (Bacchi 2009; Plumwood 1993, 2002). Plumwood distinguishes five characteristics of dualisms; backgrounding, radical exclusion, incorporation, instrumentalism and homogenisation which may be used in conjunction with each other as mechanisms to reinforce superiority or inferiority. Dualism is evident, for example in Ladson-Billings (2009) notions of conceptual whiteness and blackness where structural privilege and disadvantage are embedded in everyday discourses of conceptual whiteness. Here objects, practices and institutions associated with whiteness are normatively considered excellent and desirable; in this case universities. The converse is conceptual blackness where institutions and practices associated with blackness are normatively considered to be mediocre and less desirable. Plumwood suggests a number of mechanisms that enable dualism and the entrenchment of difference that benefits structural superiority. *Backgrounding* (also referred to as denial) entails making use of the other to service one's own needs but denying dependence on the other – what Joan Tronto (1993, 2013) refers to as 'privileged irresponsibility'. *Radical exclusion* (also referred to as hyperseparation) occurs where the differences between the inferiorised and superior groups are maximised and essentialised, and where shared qualities are minimised. This is evident in constructions of white institutions as excellent and internationally competitive whereas black institutions are considered mediocre and not internationally competitive (Robus and MacLeod 2006). These perceptions are promoted through physical and geographical separation of groups, as were achieved through the Group Areas Act during the apartheid era, which continues to impact on where people live and what resources they have access to. This also applies to access to higher education institutions in South Africa (Department of Higher Education and Training 2010). *Incorporation* defines the inferior side of the duality as inessential and the superior side as the reference point, whose qualities are the primary. The resulting misrecognition erodes opportunities for mutuality or equal relationships. *Instrumentalism* (objectification) is a form of objectification where those on the inferior side are not recognised as having their own needs. Empathy for the other is non-existent and the other is only useful in terms of meeting the needs of the dominant group. *Homogenisation* (stereotyping) occurs when differences within the inferiorised group are disregarded and they are all seen as the same. All students from a particular university or profession may be regarded as similarly inferior or superior, their differences are minimised and they are regarded as interchangeable with each other.

Therefore, in spite of formal desegregation, an informal spatial segregation between and within institutions remain in South African higher education. Given the stark dualisms evident in these normative constructions of institutions and the people that inhabit them, it is crucial to deconstruct and add complexity to commonsense understandings of difference, when considering anti-racist interventions.

Our project set out to challenge or address these dualisms by providing spaces and planning activities where students across multiple differences could come together and share everyday experiences and political histories. Even though numerous theories have been employed to discuss student perceptions of universities, Plumwood's notion of dualism highlights core essentialisms in dualistic thinking. We provided opportunities for face-to-face contact and working together, we aimed to provide opportunities for students to engage in human exchanges where the uniqueness and individuality of each person could be valued and where they would see each other as having legitimate needs. We hoped that students would be able to acknowledge their interdependence on each other's institutions and professions by recognising their value. We furthermore anticipated that the recognition may assist students in reimagining aspects of their identities in the light of their experiences of difference.

Table 1. Demographic information of all CSI students (*N* = 282).

	Year	2006	2007	2008	Total	%
Discipline	Psychology	41	14	13	68	24.1
	Social work	50	44	54	148	52.5
	Occupational therapy	N/A	44	22	66	23.4
Gender	Female	78	93	77	248	87.9
	Male	13	9	12	34	12.1
Race	African	19	30	36	85	30.1
	Coloured	43	58	35	136	48.2
	White	29	14	6	49	17.4
	Indian	0	0	2	2	0.7
	Not specified	0	0	10	10	3.6
Language	African	17	30	30	77	27.3
	Afrikaans	46	22	16	84	29.8
	English	28	50	41	119	42.2
	Other	0	0	3	3	1.1
Total		91	102	89	282	100

Methods

We embarked on a teaching and learning research project across two HEIs – the University of the Western Cape (UWC), a historically disadvantaged or black institution serving a largely working class student group (Breier 2010) and Stellenbosch University, (SU) an historically white or advantaged institution serving a largely middle class student group and across three human service disciplines in the Western Cape. The team consisted of educators in psychology, social work, occupational therapy and teaching and learning professionals. Plumwood (1993) notes that both continuity and difference have to be dealt with to overcome the dualistic dynamic. We thought about how best to provide opportunities for students to encounter each other intersubjectively by giving them opportunities to illuminate their histories, realities and their needs to attempt mutual recognition – experiencing each other as both similar and different. We used various mechanisms to do this – PLA techniques, online discussions, performances by artists and poets, critical literature, group presentations and reflective essays. We ran this course over a period of six weeks, annually, for three years (2006, 2007 and 2008) and a total of 282 students were involved over the entire period. The assessment was continuous and multi-faceted. Student engagement in small online groups of about six members, group presentations as well as reflective essays were assessed by trained facilitators. Rubrics were used in defining assessments criteria clearly (Leibowitz et al. 2012). Even though the project was not initially set up to specifically examine student perceptions of each others' and their own universities, strong views about the two universities emerged from student engagements and was deemed worthy of analysis, given the stark evidence of dualisms. In this article, we focus only on UWC and US students' constructions of their own and the other institution in a worksheet assignment which was set for them in the Community Self and Identity course. We analysed 157 students' responses to a worksheet given to them in 2007 and 2008 where two of the questions that they were asked were the following: *What did you learn about your [own] institution in your encounter at the workshop?* (our insertion) and *What did you learn about the other institution at the workshop (Stellenbosch or UWC)?*

See Table 1 below for a breakdown of the disciplines, and gender, 'race'[1] and home language of the students.

Findings and discussion

Examples of dualistic perceptions of universities

Perceptions of Stellenbosch

The most common perceptions of Stellenbosch University by UWC students are that it is a racist, Afrikaans, white, elite institution that maintains high quality and academic excellence, and where students worked very hard. The following four quotes are emblematic of many of the quotes reflecting perceptions of Stellenbosch University:

> we have heard that they are racist, and that their standards are high. We assume that most of the students are 'white'. they have a lot of resources that we do not have. They are very expensive, but the quality is good (UWC black social work female student, 2008).

> I know that Stellenbosch is regarded as an elite university; one of the best in this country. It also consists of mainly white Afrikaans students (UWC black social work female student, 2008)

> They are hardworking, racist and do more work than us, They have far more resources than we do (UWC black social work female student, 2008).

> It seems that Stellenbosch University is attended by predominantly White people mostly Afrikaans-speaking. Although there are other races and cultures that attend, it seems that each culture tends to group together in friendship circles with few cross-culture "cliques"(UWC black occupational therapy female, 2007).

These quotes all use incorporation and radical exclusion in students' dualistic perceptions of the Stellenbosch University. While not always flattering of SU (for example the reference to racism), there is a deferential status ascribed to the university suggesting incorporation. SU is regarded as excellent and students are perceived to be hardworking. In contrast the perception is created that students at UWC are not hardworking which implicitly reinforces negative perceptions of UWC. Similarly, structural issues such as access to resources and historical material privilege are mentioned in relation to Stellenbosch University being well resourced when compared to UWC. The maintenance of historical material privilege in historically white universities, when compared to historically black universities in contemporary South Africa, is consistently debated with calls to national government for reforms in funding policies for higher education that recognise and address current inequity (Bozalek and Boughey 2012). Radical exclusion is also used as a mechanism to emphasise dualisms. Differences between the institutions are highlighted with little emphasis on commonalities, whether perceived or real.

Similarly, perceptions of UWC were coloured by dualisms from both UWC and SU students. Common perceptions of UWC were both complementary and denigrating, these two perspectives often co-existing within the same quote. Perceptions about UWC as offering a poor quality education and events being disorganised were common. Students also thought that UWC offered students an opportunity to be educated in a culturally diverse setting and that the fact that it is an English medium university made the institution more universally accessible. The question of language is central to current debates on inclusion and enabling an anti-racist University in South Africa. At a university like Stellenbosch University, Afrikaans is the primary language of communication, alongside English. The language policy states that Stellenbosch aims to be a multi-lingual university. In practice, Afrikaans is dominant and still ensures that the university remains predominantly white and retards racial integration as Afrikaans as a language of higher education is highly racialised. This is an ongoing and complex debate and it is beyond the scope of this article to engage fully with the issue of language in higher education. Suffice to say that it is not as much the language as it is white Afrikaans cultural domination that accompanies the language, that diverts even black first language Afrikaans speakers, to an English medium university, such as UWC.

Perceptions of UWC

The UWC students were conscious of their university not always being regarded as organised and as an institution of quality and indicated their feelings of relief when the workshop was perceived to be conducted in a professional way:

> I learned that I can feel proud to be studying at the University of the Western Cape. It is often the perception from students from other Universities that the education that we receive at UWC can't possibly be as good as other Universities. During the first workshop at our University I felt that our students and lecturers were very professional and well spoken. I particularly felt proud when (lecturer's name) gave her lecture. I thought that it was very good, easy to understand and clear.

> The day was also well organized (something that can't always be said of UWC) and everything was done to make the other students and the guest lecturers feel at home. The fact that there was something to eat and drink was also very thoughtful. I feel proud to be part of the multi-cultural environment that we have at UWC and feel that this is an advantage that other students do not necessarily have. It is a great learning experience to be exposed to so many different cultures and different ways of thinking. The fact that we are an English medium University also helps to make us more universal. (black, female, 2007, UWC, SW).

UWC students also commonly and proudly described their university as multi-cultural and as an English medium university, emphasising its universality. In this moment of pride, UWC students use radical exclusion to highlight the differences between them and Stellenbosch University, suggesting that Stellenbosch, as an institution, is monocultural, Afrikaans, and local with little global appeal as a perceived Afrikaans language institution. Their quality judgements and pride are complex as their radical exclusion of UWC draws on false but commonsense internalised racialised discourses about white competence and black incompetence to symbolise institutional quality. Drawing on racialised discourses is common when making decisions or judgements in everyday life in South Africa (Stevens, Duncan, and Hook 2013). At the same time, the discourse of multi-culturalism is drawn on with pride and is viewed as beneficial by students. The notion of multi-culturalism is controversial and cannot be discussed fully here. However, the students use multi-cultur-alism as a positive institutional attribute. This is interesting because UWC is not racially diverse as a predominantly black institution, but is diverse in terms of religious, ethnic and language diversity. Students draw on common racialised euphemisms of culture as black-ness when they use multi-culturalism as a positive institutional attribute. They believe that multi-culturalism will help them to work in integrated work settings, that is, work settings where there are also black people, after qualification (Carolissen 2012).

On the other hand, many Stellenbosch students indicated a more derisory attitude towards UWC and also towards the higher education institution next door to UWC – the Cape Peninsula University of Technology (CPUT), also indicating that they did not know much about the institution:

I know that it's in Bellville and that its acronym is also known as the University of Wild Coloureds[2] and that Trevor Manuel[3] used to study there (Stellenbosch white psychology student; 2008)

Honestly I don't know much about UWC. But I would love to learn more. I think it's situated somewhere in the Bellville area? (White psychology Stellenbosch student; 2008)

… a university where tuition was exorbitantly high, a well resourced university, and one where residential students enjoyed more space than our 3 × 3 dormitory rooms had to offer. Thus, persons who gained entry to an institution such as Stellenbosch not only had to meet strict academic standards, a coherent feature shared by our university, but also had to pay a huge sum of money in registration and tuition fees, a parameter I regard as flagrantly exclusionary (UWC black female social work student).

In these quotes, students use radical exclusion by claiming little knowledge about the other institution, minimising their apparent interest in the institution. The use of the terminology 'Wild Coloureds' in the first quote is also indicative of radical exclusion where the other is inferiorised, and differences between the students at UWC are maximised and essentialised. There is also a level of homogenisation in this response as all 'coloured' people associated with UWC are regarded in the same light – from 'Wild Coloureds' to the then Minister of Finance. In the last quote by a UWC student, the relative wealth and resources such as space which SU has access to and the high fees which are charged, make the institution, as viewed by the student, deny access to those who do not have resources. The class-based denial of access to higher education, because of material privilege, is one of the core issues highlighted in the #Feesmustfall student protest movement that gained momentum in South Africa and internationally, during 2015.

UWC is in Bellville, next to CAPUT (Pentech)[4]. It is a University that has much cheaper fees than Stellenbosch. I would have considered going to UWC had their degrees been recognised internationally. There are majority 'black' and 'coloured' students attending UWC (White Stellenbosch Psychology student, 2008).

In the quote above the mechanisms of incorporation are evident – as UWC is placed together with CPUT which is referred to in a derogatory manner (CAPUT). UWC is regarded with scepticism, as not having the capacity to offer degrees which are internationally competitive, and as accommodating mainly black students (radical exclusion) by implicitly highlighting the differences between this institution and the one that the student is from (SU). It is crucial to highlight that student mobility and employability in South Africa based on the institution where the degree was obtained is a complex one. UWC is one of the top historically black institutions in South Africa in terms of teaching and learning and research productivity. Degrees from UWC are internationally recognised but it is a common experience in South Africa, though not well researched, that the everyday discourse of white competence and black incompetence transfer to institutions like universities and therefore impacts on choice of university, for those who can choose which university to attend. These discourses also impact on potential employers and most black students are keenly aware of the intersection of the racialised nature of employment opportunities and the university where the degree was obtained. The intersection of race, class and institutional affiliation adds more complexity too as black graduates from historically white universities appear to be preferred to black graduates from historically black universities. However, white graduates, and especially white women, are given preferential treatment in appointments and promotions as they are the greatest beneficiaries of affirmative action policies (Department of Labour 2014).

The analysis of students' quotes using Plumwood's characteristics of dualism in this section of the paper shows how both UWC and Stellenbosch students hold dualistic and often derogatory views of their own (in the case of UWC) and the other institution but also how complex deeply racialised discourses and practices matter in terms of institutional choice for study and its intersections with race, class and gender.

A very small minority of students' views about the other institution were not changed. In the case of two students, their original beliefs were entrenched. They believed that Stellenbosch University was more organised than UWC, that mostly white, Afrikaans-speaking students attended Stellenbosch University, irrespective of their experiences to the contrary. Three students raised new issues at the end of the module that were not initially highlighted. These ranged from learning about the political and activist history of UWC, of which they were not aware, learning that UWC is geographically more isolated from shops than Stellenbosch University, that the majority of UWC students use public transport to travel to university and that UWC students' boyfriends were typically not university students. This was apparently not the case for Stellenbosch students.

The majority of students' dualistic perceptions of institutions were changed during pedagogical encounters. These are discussed below.

Addressing dualisms

Many students challenged their own dualistic perceptions through their participation in the course. They were asked how they viewed their own and the other institution at the end of the course and re-evaluated their dualistic perceptions based on experiences with students from the other university. Some students' feedback suggested a complete reframing of their views while others indicated a more nuanced reframing of views. The following quotes show how students changed their views about Stellenbosch University being an exclusively Afrikaans university and that UWC was a 'place of quality, a place to grow'

> I always thought it was an Afrikaans institution and that all of their lectures are given in Afrikaans but I came to know that there are also English lectures given since the girl in the group was an exchange student and she can only speak English (UWC, SW, 2007, black F)

The quote suggests that the student's assumption about language at Stellenbosch was reframed when she met a student who was studying at SU and could speak only English. Similarly, the quote below suggests a complete reframing in the students' interactions with students from Stellenbosch University

> I got a different view from the students at the Stellenbosch university as it was usually described as a very racist university. I discovered that stereotypes can influence your perceptions about something. By engaging with the students I realised that we are all individuals and that we should make decisions on what we experience. It was wonderful engaging with the students from the different departments and Stellenbosch University. It also taught me about diversity and how important relationship building is between students who have to one day go out into the field. (Social work, black female, 2007, UWC)

Some students such as the SU student in the quote below reframed their perceptions of UWC based on his lived experience during the course. It is interesting though how this student, as an ex-UWC student appeared to begrudgingly accede to the possibility that UWC should be viewed as an excellent institution as well. When complimenting UWC, he uses a very derogatory term 'bush' that was a very popular derogatory name for UWC among some sectors, especially prior to the 1980s (Lalu and Murray 2012)

> Well I learnt that UWC is not as bad as I had thought it was, last year I was there and I had no idea that their KEWL[5] programme was this advanced. for me that was a shock for in my years having been at bush I had developed a negative attitude towards the institution, needless to say these thoughts were completely removed upon this workshop, because for the first time UWC showed me that indeed it's a quality place of education. for this it gets two thumbs up (Stellenbosch, Psychology, black, M, 2007).

The following quote also shows how a UWC student recognises how distancing and inferiorisation are central to his constructions of UWC and Stellenbosch but that his views are 'unfounded'. He does though, identify differences in interactional dynamics between students from SU and UWC, thus suggesting that there are instances of dualistic thinking that are unwarranted but others that are not

> I'm sure that there's a tacit (though erroneous) belief (which I too am guilty of) amongst my classmates that institutions such as UCT and Stellenbosch are generally regarded as superior to UWC, which in hindsight is a totally unfounded notion. To be quite honest, the initial thought of being grouped with students from Stellenbosch was a bit frightening. However, the fear of being overwhelmed by Stellenbosch and thoughts of "Little old us (UWC Social Work students)" soon dissipated after initial greetings and getting to know one another a bit better.…….. For example, after sharing our community maps with one another, it became clear that the White students in our group were significantly more privileged than the black and 'Coloured' students. The White students also seemed to be more individualistic or independent than the rest of the students at our table, where fond references to family and community were more common amongst the black and Coloured students as opposed to the White students, who both lived independently of their parents (UWC, Psychology, black,M, 2007).

This nuanced reframing highlighted in the quote above is further supported by another UWC student

> My perception however changed as the initial workshop progressed. Students from our institution, particularly the social work students contributed quite meaningfully throughout the workshop and also seemed more keen to participate during group activities and when providing feedback. Moreover, a striking feature of the students of my institution and that of Stellenbosch, was that we seemed to be a more culturally and racially diverse group than they were. This feature I believe led to our feeling more comfortable engaging members from both my own group as well as the larger group, as the students from UWC were accustomed to working in groups where members varied along lines of race, language, culture, and gender … (UWC black female social work student).

The data suggest that students initially enacted dualisms through the mechanisms of radical exclusion, incorporation and homogenisation that exist in a complex structural intersectional web of race, language, class and institutional affiliation. There was little evidence of backgrounding and instrumentalism in the data that we presented in this paper, though these mechanisms were evident in other data collected during the broader study. Previous studies also primarily indicated radical exclusion as commonplace among university students (Robus and MacLeod 2006; Steyn and van Zyl 2001).

Radical exclusion was addressed by bringing students together who had not interacted due to the continuing legacy of apartheid and its geographical separation. This is still being played out in higher education contexts and in everyday social encounters. By deliberately organising groups of differently positioned (HEI and social identities) students together, the course designers provided opportunities for students to encounter 'the other'. The use of Participatory Learning and Action (PLA) techniques was necessary to facilitate dialogue and engagement about issues of difference among students. When doing drawings about their experiences and then discussing these experiences, students were able to participate in dialogue on an equal footing. The fact that students, due to the apartheid geographical divides, were physically separated from each other, initially supported well-entrenched stereotypes about each others' institutions. Even though all students are still experiencing the legacy and the associated materiality of Apartheid, the visceral experience of visiting each others' institutions in this curriculum renewal project minimised the mechanism of radical exclusion.

The mechanism of incorporation was also addressed as all students were able to interrogate their own identities and assumptions through hearing about the experiences of marginalisation of the racialised, gendered, classed and institutionalised other. This provided the opportunity to develop a consciousness of relationships of privilege and domination which still affect relationships between privileged and marginalised positionalities of students in the South African context. Student interrogation of their personal and institutional identities helped them to reframe some of their stereotypical assumptions in relation to their own and the other institution.

Through face-to-face encounters with each other and hearing about the effects of past and present inequalities and injustices on the lives of their peers, homogenisation as a mechanism, was also challenged. All students were given the opportunity to witness the uniqueness and individuality of those who have been othered. They therefore had the opportunity to see 'the other' as human, non-homogenous and as individuals with different life and political trajectories. In their responses to hearing the stories of their peers, it was apparent that those who were privileged realised, perhaps for the first time, the complexities of the South African situation with its apartheid legacy which still affects the lives of their fellow students. They became acutely aware of the needs of people they had hitherto been oblivious of. Marginalised students also realised that those in privileged positions did not exemplify good and perfect qualities only, but were individuals who also had vulnerabilities.

Conclusion

All previous studies concerned with student perceptions of South African HE institutions are located within specific cases of particular South African institutions. This study draws on inter-institutional student perceptions derived from dialogue and pedagogical engagement, located within ongoing complex dominant discourses about HE. We have argued the importance of Plumwood's notion of dualisms as an appropriate conceptual framework for this study and that both black and white students from these differently placed HEIs initially maintained their distance from each other institutionally. They did this largely through the mechanisms of radical exclusion, incorporation and homogenisation. They displayed racialised and classed constructions of their own and the other institution, pointing to the internalisation and pervasiveness of hegemonic discourses about higher education institutions. Stellenbosch, the historically white institution was consistently associated with excellence, quality and international stature, while this was not initially a perception of UWC. This finding echoes the work of Robus and MacCleod (2006), Steyn and van Zyl (2001) and Cross and Johnson (2008). Following the collaborative educational experience, many, though not all students, were able to reconstitute for themselves the way in which their institutions and own subjectivities were constructed in relation to their own and the other institution. The fact that a small minority of students were not able to gain insight into their positionalities in relation to others is not a shortcoming of the course. It is a deeply political and emotional question central to a pedagogy of discomfort (Boler 1999; Boler and Zembylas 2003) that we have used as transformative pedagogical strategy in this course. The core assumption of a pedagogy of discomfort is that many of us (teachers and students) are emotionally invested in protecting ourselves from that which we do not want to know and which makes us uncomfortable. Pedagogical interventions using this strategy require difficult emotional labour from facilitators and students. Facilitators need to develop a facilitation style that affirms curiosity about silenced and emotionally charged conversations about privilege and oppression in the context of the curriculum (see Bozalek et al. 2010 for detailed description of the methodologies). In order to disrupt dominant discourses, it is important to engage with hegemonic views in the context of transformative pedagogical practices as opposed to avoiding its importance in the classroom. Many students who had done this course, acknowledged in an 18 months post completion evaluation, that the benefits of engagement across multiple boundaries and insight into their own positionalities, were much clearer after they had left the module, had opportunities for further reflection and had entered the workplace (Carolissen 2012). In order to disrupt dominant discourses, it is important to provide collaborative spaces to engage with student stereotypes about institutions that, in divided and unequal societies like South Africa, are inherently racialised, gendered and classed, and part of institutional cultures. These kinds of engagements are crucial for cultivating democratic processes of citizenship and transformation of institutional cultures in HE, an issue of global concern.

Notes

1. The apartheid racial classifications of African, Coloured, White and Indian are still currently referred to in South Africa in official documentation, although it is generally recognised that these are contested and constructed categories.
2. UWC, the University of the Western Cape is derogatorily referred to as the University of the Wild Coloureds.
3. The ex-Minister of Finance from 1996–2009 during the presidencies of Mandela, Mbeki and Motlante.
4. The Cape Peninsula University of Technology (CPUT) is referred to as CAPUT (from the German word 'kaputt'), in a derogatory manner, which means 'useless, broken, dead, no longer functioning' in colloquial terms. It was known as the Peninsula Technikon, shortened to Pentech before 1994. Before 1994, technikons were perceived to offer technical, applied professional training and issued diplomas. After 1994, the role of technikons changed so that they became Universities of Technology that conferred degrees.
5. KEWL - is the e-learning platform that was being used by UWC at the time.

Disclosure statement

No potential conflict of interest was reported by the authors.

References

Ancis, J. R., W. E. Sedlacek, and J. J. Mohr. 2000. "Students' Perceptions of Campus Cultural Climate by Race." *Journal of Counselling and Development* 78 (2): 180–185.

Bacchi, C. 2009. *Analysing Policy: What's the Problem Represented to Be?*. New South Wales: Pearson.

Bartram, B., and C. Bailey. 2009. "Different Students, Same Difference?: A Comparison of UK and International Students' Understandings of 'Effective Teaching'." *Active Learning in Higher Education* 10 (2): 172–184.

Boler, M. 1999. *Feeling Power: Emotions and Education*. New York: Routledge.

Boler, M., and M. Zembylas. 2003. "Discomforting Truths: The Emotional Terrain of Understanding Difference." In *Pedagogies of Difference: Rethinking Education for Social Change*, edited by P. Trifonas, 110–136. New York: Routledge.

Bourdieu, P. 1986. "The Forms of Capital." In *Handbook of Theory and Research for the Sociology of Education*, edited by J. Richardson, 241–258. New York: Greenwood Press.

Bozalek, V., and C. Boughey. 2012. "(Mis)framing Higher Education in South Africa." *Social Policy & Administration* 46: 688–703.

Bozalek, V., R. Carolissen, L. Nicholls, B. Leibowitz, L. Swartz, and P. Rohleder. 2010. "Engaging with Difference in Higher Education Through Collaborative Inter-Institutional Pedagogical Practices." *South African Journal of Higher Education* 24 (6): 1023–1037.

Breier, M. 2010. "Dropout or Stop out at the University of the Western Cape?" In *Student Retention and Graduate Destination: Higher Education and Labour Market Access and Success*, edited by M. Letseka, M. Cosser, M. Breier, and M. Visser, 53–65. Cape Town: HSRC Press.

Brinkworth, R., B. McCann, C. Matthews, and K. Nordström. 2009. "First Year Expectations and Experiences: Student and Teacher Perspectives." *Higher Education* 58: 157–173.

Brown, R. M., and T. W. Mazzarol. 2009. "The Importance of Institutional Image to Student Satisfaction and Loyalty Within Higher Education." *Higher Education* 58 (1): 81–95.

Burr, V. 1995. *An Introduction to Social Constructionism*. London: Routledge.

Carolissen, R. 2012. "Student experiences of the CSI module." In *Community, Self and Identity: Educating South African University Students for Citizenship*, edited by Brenda Leibowitz, Leslie Swartz, Vivienne Bozalek, Ronelle Carolissen, Lindsey Nicholls, and Poul Rohleder, 59–72. Cape Town: HSRC Press.

Cooper, D. 2015. "Social Justice and South African University Student Enrolment Data by 'Race', 1998–2012: From 'Skewed Revolution' to 'Stalled Revolution'", Higher Education Quarterly, 69 (3):237–262.

Cross, M., and B. Johnson. 2008. "Establishing a Space of Dialogue and Possibilities: Student Experience and Meaning at the University of the Witwatersrand." *South African Journal of Higher Education* 22 (2): 302–321.

Denson, N., and M. J. Chang. 2009. "Racial Diversity Matters: The Impact of Diversity-Related Student Engagement and Institutional Context." *American Educational Research Journal* 46 (2): 322–353.

Department of Education. 2008. *Report of the Ministerial Committee on Transformation and Social Cohesion and the Elimination of Discrimination in Public Higher Education Institutions*. Pretoria: Government Printers.

Department of Higher Education and Training. 2010. *Stakeholder Summit on Higher Education Transformation*. Cape Town: Cape Peninsula University of Technology.

Department of Labour 2014. Annual Report 2014/2015. Accessed 12 January 2015. http://www.labour.gov.za/DOL/downloads/documents/annual-reports/departmental-annual-reports/2015/annualreport2015_part1.pdf

Fraser, F. 1997. *Justice Interruptus – Critical Reflections on the Postsocialist Condition*. New York: Routledge.

Gergen, M., and K. J. Gergen. 2003. *Social Construction: A Reader*. London: Sage.

Hemson, C. 2006. *Teacher Education and the Challenge of Diversity in South Africa*. Pretoria: HSRC Press.

James, Richard, Kerri-Lee Krause, and Claire Jennings. 2010. *The First Year Experience in Australian Universities: Findings from 1994 to 2009*. Centre for the Study of Higher Education: University of Melbourne.

Jansen, J. 2009. *Knowledge in the Blood: Confronting Race and the Apartheid Past*. Cape Town: UCT Press.

Ladson-Billings, G. 2009. "Just What is Critical Race Theory and What's it Doing in a Nice Field Like Education?" In *Foundations of Critical Race Theory in Education*, edited by E. Taylor, D. Gillborn and G. Ladson-Billings, 17–36. London: Routledge.

Lalu, P., and N. Murray. 2012. *Becoming UWC: Reflections, Pathways and Unmaking Apartheid's Legacy*. Cape Town: Centre for Humanities Research, UWC.

Leibowitz, B., Leslie Swartz, Vivienne Bozalek, Ronelle Carolissen, Lindsey Nicholls, and Poul Rohleder, eds. 2012. *Community, Self and Identity: Training University Students for Transformation in South Africa*. Cape Town: HSRC Press.

Leibowitz, B., Antoinette van der Merwe, and Susan Van Schalkwyk, eds. 2009. *Focus on First-year Success. Perspectives Emerging from South Africa and Beyond*. Stellenbosch: SUN MeDIA.

McKinney, C. 2004. "A Little Hard Piece of Grass in Your Shoe": Understanding Student Resistance to Critical Literacy in Post-Apartheid South Africa." *Southern African Linguistics and Applied Language Studies* 22 (1-2): 63–73.

Mckinney, C. 2005. "A Balancing Act: Ethical Dilemmas of Democratic Teaching Within Critical Pedagogy." *Educational Action Research* 13 (3): 375–392.

McKinney, C. 2007. "Caught Between the "Old" and the "New"? Talking About "Race" in a Post-Apartheid University Classroom." *Race Ethnicity and Education* 10 (2): 215–231.

Plumwood, V. 1993. *Feminism and the Mastery of Nature*. London and New York: Routledge.

Plumwood, V. 2002. *Environmental Culture: The Ecological Crisis of Reason*. London and New York: Routledge.

Robus, D., and C. MacLeod. 2006. "White Excellence and Black Failure": The Reproduction of Racialised Higher Education in Everyday Talk." *South African Journal of Psychology* 36 (3): 463–480.

Ruohoniemi, M., and S. Lindblom-Ylänne. 2008. "Students' Experiences Concerning Course Workload and Factors Enhancing and Impeding Their Learning – A Useful Resource for Quality Enhancement in Teaching and Curriculum Planning." *International Journal for Academic Development* 14 (1): 69–81.

Soudien, C. 2012. *Realising the Dream: Unlearning the Logic of Race in the South African School*. Pretoria: HSRC Press.

Stevens, G., Norman Duncan, and Derrick Hook. 2013. *Race, Memory and the Apartheid Archive*. Johannesburg: Wits University Press.

Steyn, M., and M. van Zyl. 2001. *Like that Statue at Jammie Stairs: Student Perceptions and Experiences of Institutional Culture at the University of Cape Town. Institute for Intercultural and Diversity Studies of South Africa*. Cape Town: University of Cape Town.

Stoughton, E. H., and C. Sivertson. 2005. "Communicating Across Cultures: Discursive Challenges and Racial Identity Formation in Narratives of Middle School Students." *Race Ethnicity and Education* 8 (3): 277–295.

Tronto, J. 1993. *Moral Boundaries: A political Argument for an Ethic of Care*. New York: Routledge.

Tronto, J. 2013. *Caring Democracy: Markets, Equality, and Justice*. New York: New York University Press.

Higher education, de-centred subjectivities and the emergence of a pedagogical self among Black and Muslim students

Pete Harris, Chris Haywood and Mairtin Mac an Ghaill

ABSTRACT

This article explores late modern Black and Muslim young men's and women's experiences of higher education. Carrying out qualitative research with 14 male and female young people, these students claimed that their Youth and Community Work course at their university made available an alternative representational space, enabling them to develop a major transformation of their sense of identity and self. In deploying the term *pedagogical self*, we are attempting to capture their naming pedagogy as central, in their terms, to the 'reinvention of their selves'. We conclude by suggesting that our research participants' narratives are located within an exploration of late modern identity and the self in higher education. In turn, this enables us to reflect on a generational shift in meanings around racialisation and difference in thinking about the future of higher education in Britain.

Introduction

James: I'm glad I've come on this course and I really want to finish with a high grade and just prove to myself that I could and be a bit of a role model to people around me that will see me graduate and see the possibilities for themselves and hopefully to show other people it's possible, symbolic capital. Have my little picture on the wall of me graduating, my certificate. I guess it's going to help me grow in my practice, the kind of support I can offer to young people as well.

Hameeda: Looking back, I feel for us on our course we've made something of ourselves.... something important. I never liked school, it was bad but I always knew I liked learning. And here it all came together, all the discussions, all the arguments and everything. For the first time in our lives, we believed in ourselves and each other and it's great... It's like we came into uni and now we're going out a completely different person. And we can pass this on in our professional practice.

Currently, for black and Muslim young men and women, much political, media and academic commentary serves to re-inscribe them as a major social problem for the state. This representation builds on the histories of British higher education that are littered with discarded images of diasporian groups, marked by homogenisation, dysfunction, failure, lack of voice, marginality and invisibility (Bird et al. 1992). In an earlier period, questions about 'anti-racist' practice across educational sectors in response to pervasive racial discrimination revolved around collective mobilisation against institutional racism (Law 1996). More recently, structured inequality has been reframed as individual failure as the neoliberal discourses of individual responsibility and choice drive de-racialised subjects in the cultivation of knowledge and skills deemed to be in global market demand (Haywood and Mac an Ghaill 2013). This is occurring within the context of a performative-based higher education system, at a time of the emergence of an assertive English nationalism involving a forging of a renewed British identity and a European-wide political questioning about state-led multiculturalism and a shift to securitisation (Miah 2015).

This paper seeks to build on previous work on black and Muslim student experience in higher education (Bagguley and Hussain 2014; Burke et al. 2012). Critically engaging with dominant representations of minority ethnic failure as individual choice, we draw upon a qualitative analysis of student narratives to explore a changing cultural condition that is inhabited by a younger generation of British-born, working-class, black and Muslim students (of Pakistani and Bangladeshi heritage). Against a pessimistic trend in the literature concerning a lack of promotion of ethnic/racial equality, the students in this study, as illustrated above by James and Hameeda, claimed that their Youth and Community Work course at a post 1992 University in the Midlands made available an alternative representational space, enabling them to develop a major transformation of their sense of identity and self. Much writing in the literature addressing questions of late modern identity and the self in education operates within an identity politics framework and our research participants generally shared this perspective (Mirza 2015). However, in deploying the term *pedagogical self*, we are attempting to capture their naming pedagogy as central, in their terms, to the 'reinvention of their selves'. The students referred to a range of themes that might be identified as constituting the pedagogical self that included: the central role of the lecturers; dialogical encounters with peers and inter-subjective recognition of the self in the 'other'; earlier educational experiences and a new (academic) literacy to name past and current experiences.

The paper begins with a discussion of our methodological approach, followed by a section on the literature and current empirical studies we drew upon exploring late modern identity and the self in higher education, as a context for setting out our research participants' narratives. This enables us in a final section to address reflections on a generational shift in meanings around racialisation in thinking about the future of higher education in Britain.

Making methodological sense: young men and women's narratives

The Youth and Community Work course is grounded in a mix of theoretical discourses, most notably Paulo Freire (1972) notion of 'dialogical' education. The aims of the course are to: equip students with the knowledge, understanding and skills required to achieve professional status in Youth and Community Work and to be a reflective practitioner; ensure that students are able to base their professional practice on a systematic understanding of the key aspects of youth and community work, and be able to apply this in different contexts; promote the development of key transferable skills necessary for employment including: the exercise of initiative and personal responsibility, decision-making in complex and unpredictable contexts, the ability to work as a member of a team and autonomously, and the learning ability needed to undertake appropriate further training; ensure a supportive learning environment which develops students' confidence and intellectual curiosity; and to enable students to adjust to the demands of learning at HE level and function in a modern professional environment including acquiring information and digital literacy. Operating as an applied, vocational course, students also learn experientially in the field often in local communities affected by poverty and social exclusion. Acting as gatekeepers to a 'profession', Youth and Community Work courses traditionally embrace a 'community of practice' model (Lave and Wenger 1991), where students learn experientially in the field under practice supervisors, who enable students to make links between practice and theory.

For this small-scale research project, we carried out unstructured interviews with eight black (five female and three male) students and six Muslim (two female and four male) students. The interviews lasted between 60 and 90 min and the students chose the location where they felt most comfortable to be interviewed within the institution. Throughout our conversations with the students, questions of contested understandings, interpretations and meanings were central, operating within specific higher education institutional power/ knowledge configurations. It is important to note that diasporian social groups are highly diverse and as a qualitative and explorative study, the article does not seek inductive validity by suggesting that the participants represent the experiences of the broader diasporian groups of the general population. Instead, as Crouch and McKenzie (2006, 493) argue:

> Rather than being systematically selected instances of specific categories of attitudes and responses, here respondents embody and represent meaningful experience-structure links. Put differently, our respondents are 'cases', or instances of states, rather than (just) individuals who are bearers of certain designated properties (or 'variables').

The exploration of the students' meaningful experiences was a key objective of the research design. The interviews were supplemented by a range of other research strategies that included observations and informal conversations, informed by our wider critical ethnography on the impact of globally inflected change upon the local formation of diasporic younger generation's subjectivity and identity (Appadurai 1991; Mac an Ghaill and Haywood 2014). The data were subject to thematic analysis (Braun and Clarke 2006) that enabled us to explore '…the underlying ideas, constructions, and discourses that shape or inform the semantic content of the data' (Ussher et al. 2013, 3). The subsequent analysis was taken back to the students themselves not simply as a form of 'face validity' but also as a way of exploring the practical and political implications of the findings. All interviews throughout the study were anonymised and the research participants were given pseudonyms.

Higher education, identity formation and the production of a pedagogical self

Bernstein's (2000) sociocultural theory of education enables us to make a fundamental critique of neoliberal discourses of British teaching and learning, with its emphasis on an overly individualised entrepreneurial self. More specifically, the key concept of pedagogy opens up a broader understanding of teaching. Bernstein's work has had a major influence on critical educationalists over the last few decades. This is most recently illustrated in Burke et al's (2012) text in which they suggest: 'We understand pedagogies as lived, relational and embodied practices in higher education. Although hegemonic discourses at play in education policy construct largely instrumentalized notions of teaching and learning, the dynamics, relations and experiences of teaching and learning are intimately tied to the re/ production of particular identity formations and ways of being a university student and teacher' (9). Bernstein's work enables us to begin to formulate a concept of the 'pedagogical self' with its suggested shift from teaching as a simple practice of transmission of knowledge from teacher to student to pedagogy as a dynamic multidimensional practice of critical engagement with and among peers and lecturers, that resonates with our students' current institutional experience.

At a time of the demise of structuralist theories, Bourdieu (1997) continues to be analytically central to understanding the (class) reproduction of education. More particularly, he is one of the major influences on 'capital' being used as an analytical framework in explaining the relationship between (in)equality, social mobility and higher education among minority ethnic groups. Within a British context, Modood (2012) has established his own analytical approach, in which social capital and ethnic capital have been key features, in exploring South Asian/Muslim experiences of higher education. Modood's work, alongside a range of other recent researchers, including Basit (2013) and Crozier and Davies (2006) carrying out studies on educational, social and ethnic forms of capital have made a major contribution in challenging earlier pathologised representations, especially of young British Muslim/ South Asian women. However, in relation to the young men and women with whom we were working, these theories tended to underplay generationally specific understandings of subjectivity, identity formation and the self, in other words processes of subjectification (Crozier and Davies 2006).

It is important to explore the generational specificity and how the pedagogical-self interplays with young people's subjectivities. Bagguley and Hussain (2014) in critiquing social capital theory, claim that the concept reflexivity provides such a generationally specific understanding of young people's experiences of higher education at a time of rapid shifting change with reference to equality, social mobility and differential subjective reflexivity. Drawing upon Archer (2007, 4), they write:

> She suggests that reflexivity is characteristic of all individuals who hold internal dialogues mediating between their social circumstances and their actions. This explains how structural constraints and enablements operate through human agency. People's course of actions are products of their reflexive deliberations and their subjectively determined personal projects in relation to the social and cultural circumstances objectively confronting them. Some of these objective circumstances will be known, but others will be unacknowledged conditions of action. It is enough to know that the unacknowledged conditions are there, but not the details of their operation, nor do they have to be 'internalised'.

Much of the theoretical legacy of social capital theory continues to appear to be derivative of a first-generation migrant experience. For Bagguley and Hussain (2014), 3, Archer's work provides a more productive framework than social capital theory in explaining 'changes in educational outcomes' within communities (3). They are particularly interested in South Asian women's subjectivity, agency, and social mobility, referring to them as reflexive commentators, whose social practices have much in common with the students in our study.

For our students in their discussion of academic representations of higher education, the ongoing narrative (and research) of the post-war immigration of South Asian/Muslim and black communities is still being told in an older generationally specific language of race and empire that is not able to grasp the specificities of a younger emerging inter-ethnic and inter-religious social relationships and their engagement with a different de-centred racial semantics, as Chhaya illustrates here.

> Chhaya: You look at unis and they're full of black and Asian people, especially young Muslim women, like lots of other young women, they've changed what unis look like. They don't want to keep hearing about failure. We need to think about how to build on this success to break the glass ceiling at work. These young women, this younger generation, have a different way of looking at things, different experiences, different meanings, and I think they will do it.

Post-colonial writers have made an important theoretical contribution in providing new frameworks that address the issues raised by the students. These writers enable us to move beyond the reductive black/white colonial paradigm, to make sense of new ethnicities, decentred subjectivities and syncretic cultures around the politics of race and nation (Brah, Mary, and Mac an Ghaill 1999; Hall 1992). Within the context of higher education, a main claim of this work on racial difference is that we cannot simply read off social relations from fixed oppositional categories of blacks and whites, marked by ethnic boundedness, fixity and social separation, in making sense of students' pedagogical experience. We suggest that an empirically based critical analysis located within higher education, enables us to conceptualise a response to emerging identifications and de-centred subjectivities among students by viewing them as a set of narratives of 'self-production' that are dispersed through a multiplicity of power relations that are actively secured. As Kevin illustrates below, the students suggested this as constituting a generational difference with their parents' experience (see McLean 2006).

> Kevin: I think that most theories about black people going onto to university, it's about our parents' generation, about people not born in Britain. It's all that deficit ideas about us, about oppression, a 1970s model….but now we need to look at how we are successful, how people are getting through.

The students are not simply talking about increasing numbers of black and Muslim/South Asian students now in higher education but significantly are making reference to their own educational biographies. As we suggested in the introduction, much writing in the literature addressing questions of late modern identity and the self in education speaks of multiple identities and selves in terms of subjects inhabiting the major social categories of gender, sexuality, ethnicity or class, and most recently religion (Mirza 2015). The students inhabited educational trajectories that were embedded in late modern understandings of cultural difference at individual and collective levels. However, in deploying the term *pedagogical self*, we are attempting to capture the students naming pedagogy as central to the 'reinvention of their selves' and accompanying de-centred subjectivities, contributing to their self-recognition of their *'becoming someone'*. Foucault's (1978) earlier work on the production of identity formation and selfhood has been influential in critical educational research, although at times it tended to emphasise the regulation and disciplining of the subject. In his later work, Foucault has suggested exploring identity as a technology of the self, where subjectivity is a socio-historical formation of dispersed institutional arenas of power (Foucault 1988). This work develops the notion of agency, enabling us to explore more sensitively the specific historical dynamics of the production of complex and diverse racialised/ethnic subjectivities (Bagguley and Hussain 2014). One of the main concepts adopted by recent theorists is the notion of the techniques of the self, that is modern forms of managing/producing the self. Foucault speaks of four types of techniques 'that he says agents practise on themselves to make themselves into the persons they want to be' (Martin, Huck, and Patrick 1988, 18). They are: technologies of production, which permit us to produce, transform or manipulate things; technologies of power, which determine the conduct of individuals; technologies of sign systems, which permit us to use symbols of signification; 'technologies of the self, which permit individuals to effect by their own means a number of operations on their own bodies and souls, thoughts, conduct, and way of being so as to transform themselves in order to attain a certain state of happiness, purity, wisdom, perfection or immortality' (Martin, Huck, and Patrick 1988, 18). As Nixon (1997, 323) maintains, of particular significance here is that: 'Foucault's comments on 'practices of the self' open up the possibility of conceptualising the articulation of concrete individuals to particular representations as performance based upon the citing and reiteration of discursive norms; a performance in which the formal positions of subjectivity are inhabited through specific practices or techniques'.

We now move to use the above theoretical insights in analysing our research participants' narratives about their recent experiences of higher education. As we indicated in the introduction, reading through their narratives a range of interconnecting themes can be identified with reference to the constitution of the pedagogical self, that included: the central role of the lecturers; dialogical encounters with peers and inter-subjective recognition of the self in the 'other'; earlier educational experiences and a new (academic) literacy to name past and current experiences.

Central role of lecturers

A major theme that the students returned to throughout the fieldwork was the critical importance of the lecturers on the youth and community course in the co-production of a pedagogical self (Bhopal and Danaher 2013).

Leonie: How did it all start? The biggest thing was the lecturers believing in us. Yes, even when there were lots of tensions and angers and confusion and silences in the discussions, they stuck with us and believed in us. Then, we began to believe in ourselves. It was like when we came into the lecture room a light went on and we became new people.

Smita: It seemed like the first time we could really discuss important things. At the time, it seemed it was all about learning new theories and new ways of thinking and arguing. A big thing was listening to people and their ideas that you did not agree with…. But the biggest thing we were learning was about ourselves. It's strange, we've all trained as youth and community workers and we're in situations at work where we never stop discussing things. So, what was so different about this place with these people and these lecturers? Because this was a teaching, a learning place.

Sukhdip: It seemed like the lecturers were transforming us at the start, by the end of the first year, you could see, we were changing ourselves in important ways. It was like being in a play, sometimes the lecturers were the directors, sometimes we were directing ourselves. I think even the lecturers were learning things from us…. Imagine going back to school and telling the teachers what was happening to their little pupils that they wrote off all those years ago.

Of central significance for the students on the course was that they were encouraged to think critically at a time when they were intrinsically motivated to do so; with each other, and with lecturers. This was achieved by bringing theory into the visceral, embodied experience of lectures and small tutor groups and by working to bring tacit, unconscious processes into a learnable, theoretical framework. In this sense, the process of self-development became part of the pedagogical purpose of the course. The use of 'generative themes' (Freire 1972) that emerged from the students' own reality and were raised by them were considered central to this process. The students were highly motivated to examine how they were positioned within their social world, when engaged in an educative process that began with themes drawn from their immediate, concrete reality. This allowed for both seemingly trivial and significant aspects of their own lives to be first discovered, named, and then imbued with meaning. In turn this enabled them to act more autonomously and in ways that precipitated their personal development and change within their changing social reality.

Dialogical encounters with peers and inter-subjective recognition of the 'other'

Much of the earlier work both in anti-racist theory and social capital theory with reference to Black and South Asian/Muslim students' educational experiences have under-played their complex identifications, affiliations, investments and positionings. More specifically, within the location of higher education, we know little about the complex processes of subjectivity and accompanying processes of subjectification, inter-subjectivities and social biographies, complex investments/affiliations and the occupying of multiple and diverse identifications (Benjamin 1998). For our students, equally of critical significance as their lecturers in the development of the pedagogical self, was their interaction with their peers. They claimed that the university was an institutional space that provided them for the first time in their lives with an opportunity to meet a wide range of people with different life experiences with whom to engage in critical dialogue.

> Joanne: I feel there is definitely an ethos of, like…equality, and anything, any kind of oppression is challenged….there is a space for dialogue…so even if people do have views that I wouldn't like or question, I have to force myself to listen acceptingly and people are allowed to say whether it's right or wrong. We can air these topics and it feels like it's important, these aren't just side issues.

The students were on the Youth and Community Work course during a specific period that included the increasing visibility of the 'Muslim question' and the re-racialisation of Muslims within Britain and across Europe (Mac an Ghaill and Haywood 2014; Miah 2015). The critical dialogue among peers and between peers and lecturers covered a wide range of educational, political, biographical and ethical issues, including providing a unique institutional 'safe space' to discuss faith and identity. Significantly, discussions that began in lecture rooms were continued among small groups of students in various leisure spaces across the university. One of the most visible friendships to develop, in which the inter-subjective recognition of the self in the other was to the fore, was between a Muslim student and a white student, who was a former soldier.

> Iftikhar: If we talk about Dave over a period of time, with his background, with my views, I feel I can have that conversation with him because I'm assured that actually he understands better than most others where I am coming from, than anybody else….me and Dave are starting to get a better understanding of each other…. Listening to Dave, talking to him, actually seeing his physical expressions when he is talking about that stuff gives a better understanding, you know, that he's actually been through a lot. : …not crossed it, but we are on that bridge, understanding each other and I wouldn't have wanted to have left here…and not been on that bridge of conversation with Dave, and we're here now and hopefully within the next semester, we can get somewhere with that and that for me would be a huge step because I'm actually starting to understand the other side.

Some of the students were especially keen to uncover the inter-subjective, dialogical processes at work within pedagogical interactions. In line with other analyses of student experience (Lucey, Melody, and Walkerdine 2006) we see merit in making use of the work of Jessica Benjamin (1998) who is concerned with how parties within a relationship make known their own subjectivity and learn to perceive and appreciate that of the 'other' – a process she calls *recognition*. Judith Butler defines recognition as,

> A process that is engaged when subject and other understand themselves to be reflected in one another, but where this reflection does not result in a collapse of the one into the other (Butler 2004, 311–12).

We utilised this concept in relation to the narratives we gathered from students as we sought to understand their engagement in a kind of semiotics of self, through which they came to personify a set of values informed by the critical pedagogy which they were collectively creating in their undergraduate course. We incorporated a notion of reciprocity to capture the simultaneous, two-way, processes whereby students 'co-produced' with each other and tutors a sense of a new self – which we refer to as the 'pedagogical self'.

Earlier educational experiences and a new (academic) literacy to name past and current experiences

One of the most visible elements for the students, in the development of the pedagogical self, was the acquisition of a new 'academic' literacy. This manifested itself in several ways, including being able to name past and current experiences and imbue them with meanings and reflexively to articulate self-representations to themselves and others. A major institutional sign of their success was that these students, who entered university with low academic qualifications completed the course with top grades, including several of them gaining first-class degrees.

Kevin: Coming to Uni I've learnt about different theories, I guess, like, things that I felt…. frustrated with….you know, anxious about in my life. I always knew they were there, because I felt them, but couldn't really describe them. But coming here I've learnt about things that I can give a label to.

Iftikhar: You know, I think I'm working at a different….. how can I put it? This way… my reflection, the way I live now is at a different algorhythm, if I can put it that way, compared to before I came and my own understanding of myself.

Earlier in the paper, we spoke of the pedagogical self as dynamic and creative. One of the most impressive examples of this throughout the students' narratives was their reflecting on their own early educational experiences and how their new self was transforming their personal, domestic and professional lives (Mirza 2015). This is a long way from the imagined autonomous 'entrepreneurial self' that is prescribed by neoliberal educational policy with reference to social mobility and widening participation.

Khalid: In Asian families there is a big pressure for children to look after their parents when they get older….. For me, I went out to get a job early to support my family….What's different now is that I'm bringing home cultural capital rather than money. What I've got from this course affects the whole family, including my brothers and nieces….. Everyone is very proud and of course they've all supported me over the years, so really, it's their success as well.

Khalid's comments are interesting from several perspectives. First, as suggested above, the use of 'generative themes' (Freire 1972) that emerged from the students' own reality and were raised by them were of central significance to the constitution of the pedagogical self. The students were constantly questioning how they were positioned within their social world, when engaged in an educative process that began with themes drawn from their immediate, concrete reality. Such engagement resonates with Freire's notion of a 'generative theme', which he suggests is a cultural or political topic of great concern or importance to people from which discussion can be generated. These generative themes can be represented in the form of 'codifications' (words, short phrases or visual representations), which people can then step back from and decode or explore critically by regarding them objectively rather than simply experiencing them. For the students, this was a central aspect of their learning, enabling them to intervene and initiate change in society, particularly with reference to the oppressive conditions experienced by their families and communities.

Second, Khalid's comments are also interesting within the context of the recent shift in policy discourse from a focus on multiculturalism to surveillance and securitisation (Miah 2015). Such recent government discursive shifts, including the suggestion that a younger generation of Muslims need to adopt British values sees a return to an assimilationist model that was put in place for a first generation of post-war migrants. Such a model, based on a false dichotomy between (religious) tradition and (secular) modernity assumes that the latter must be embraced by young Muslims by displacing and distancing themselves from the former. As Khalid finely illustrates the students were in the process of developing late modern reflective biographies by drawing upon family collective resources and accompanying values. This more nuanced understanding of racialisation, racial positioning and subjectivity was reflected on throughout the research period and is explored in the next section with reference to generational shifts of meaning (Crozier and Davies 2006).

Making sense of generational shifting meanings of racialisation: from colonial-based race to post-colonial ethnicity

When we first read the students' narratives, we had a sense of a generational shift from a colonial-based understanding of racial inequality marked by structural discrimination and collective mobilisation against institutional racism to the politics of identity formation, marked by ethnic /religious subjects, individual consciousness and multi-culture (Hall 1992). As we re-read these narratives and in discussion with team members of the Youth and Community Work course, we began to see a more complex picture of different students combining diverse emphases on the *structural* and *identity making* aspects of Muslim and Black experiences of higher education (Mac an Ghaill 1999). However, most of the students appeared to position themselves within what might be called a late modernity sensibility, highlighting a sense of reflexivity, fragmentation, ambivalence, individuality and multiple/ shifting identities. Their thinking might be captured by Bauman's (2001) notion of *Liquid Modernity*. Thus, within the context of their course, students were developing subjectivities that were marked by 'their fragmentariness and discontinuity, narrowness of focus and purpose, shallowness of contact' (Bauman 1996, 34). In other words, their sense of identity appeared to be no longer premised on fixed identity categories enabling a multitude of discourses of religion, 'race'/ethnicity and class to temporarily cohere. Furthermore, it could be argued that their narratives suggest a space that might exist outside neoliberal discourses of individualism, choice and entrepreneurialism. The pedagogical self could be understood as an active reflexive project, where as Fadeeva and Mochizuki (2010), 250 point out: '… the transition from "solid" to "liquid" modernity has challenged individuals to find alternative ways to organise their lives, for social forms no longer have enough time to solidify and cannot serve as frames of reference for human actions'. The liquid nature of their narratives is discussed further below.

Within the context of exploring the research literature on British higher education and anti-racism, among our participants, we found a range of complex, nuanced and differentiated student narratives with reference to global, national and local histories. A sense of liquidity can be found in the generationally specific individual and collective negotiation with and resistance to multiple forms of racism and faith hate, resonant of a wider politics of culture involving overlapping territories and intertwined histories (Said 1993). For example, within the interviews some students appeared to operate with what might be referred to as a certain cultural amnesia about structuralist-based 'old racial times'. When we discussed the idea of the anti-racist university, they displayed a primary interest in interpersonal social relations, that is, interactions between staff and students and among students themselves, as primarily indicative of the presence or absence of racism.

> Lisa: We have a lot of experience of racism as youth workers and this place is definitely not racist. Just look at how people get on together…. I'd say there's a lot of respect for each other for everyone of different communities.

> Farzana: This is definitely a safe space for Muslim students…. Islamophobia is everywhere in this country at a really high level…. but not, definitely not here.

There was a general reluctance to locate racism within the structure of the institution, or more specifically, the geographical and social space in which they were now located, preferring to see it more as a function of the point or place in which they found themselves in what they referred to as 'their own biographical journey'.

> Sukhdip: I don't think it has anything to do with the institution, for me anyway, it was about the students and myself in regards to, you know, how we engage with people, everybody engages with people in different ways.

Other students incorporated earlier versions of anti-racism that informed their own generationally reflexive scripts that were spoken within more individual identity-based inflected accounts of social and cultural exclusions, with a strong sense of the interconnection of multiple categories of exclusion and the emergence of the racialisation of a wide range of social groups, including recent East European migrants and refugees and asylum seekers (Mirza 2015).

> Leonard: Over the years I've shifted from just thinking of a black perspective and all the systemic racism on blacks to thinking about others who have a lot of prejudice and stereotyping. Like the recent migrants and refugees I work with. These discussions would have been very different 20 years ago, now there is more emphasis on how racism affects individuals and all the talk we had on the course of the different identities, like a lot and big arguments about working class identities and where poor white kids fit on the map of oppressions.

There was also a smaller group who articulated an explicit commitment to an anti-racist or anti-Islamophobic politics, around a race/ colour consciousness or religious adherence.

> Farzana: On the course we discussed the deep structures of racism. It often came up. Maybe a lot of people, maybe most don't seem to be interested in that anymore. But for me, yes, it's still the same, maybe some things have improved in some ways. But mainly on the surface. What's really changed is that white people

cannot say the racist stuff openly and they feel offended by this, because for them they are just speaking what they really feel. The exception is that you can say openly what you feel about Muslims. There is so much Islamophobia now.

Michael: It's like the term racism isn't really understood anymore. Of course black and Asian people still end up at the bottom on schools, jobs, bad housing, extreme poverty, etc., etc. It's not an individual prejudice, it's systemic across society… And, then a celebrity calls a black person coloured and the person is labelled racist. It's mad. We're all celebrating diversity, when the country has never been so racially divided.

Students maintained that they were re-negotiating these understandings throughout the course. Their narratives suggested that pedagogical practices addressing contemporary racial inequality are mediated through a wide range of generationally specific meanings among black and Muslim students that have continuities and discontinuities with an earlier generation of new social movement theories of (higher) educational oppression, racial inequality and cultural/religious difference. Furthermore, in trying to make sense of their meanings, it should be noted that we know little about the specificity of the cultural resources that they can call upon as individual subjects, who have vast professional expertise as youth and community workers. It is against this background, that they, as university students, narrate their shifting and diverse meanings of (ethnic/religious) identity formation and racialisation within late modern conditions of socio-economic austerity. We attempted to make sense of our research participants' narratives by framing them in terms of their working through a tension between a (colonial based) materialist understanding of race and raced subjects and that of a post-colonial ethnicity and accompanying de-centred subjectivities that included an awareness of social majorities as ethnic/religious subjects.

Materialist explanations of racism have been a central aspect of the education/training of youth and community workers over the last few decades (Sallah and Howson 2007). Materialist refers to social movements that perceive the organisation of ethnic and racialised identities as deriving from fixed bases of social power. Such bases of social power are seen to work logically and predictably, often being illustrated through an individual's occupation of fixed hierarchical positions, such as dominant/empowered (white people) and subordinate/oppressed (black people). (Mac an Ghaill 1999). This materialist position was historically important in shifting the focus of analysis away from the assumption of ethnicity acting as a barrier to black and South Asian working-class students accessing higher education to an emphasis on the commonalities of the social and cultural reproduction of racism experienced by South Asian and black students operating across institutional life.

The materialist explanation of racism continued to have resonance for the students in this study, whose experience of post-colonial Britain involves socio-economic austerity, increasing inequalities, regional socio-economic disparities and the success of UKIP in local and European elections providing evidence of an emerging new nationalism and an accompanying re-racialisation of diasporian social groups, which has become highly explicit in the period after the Brexit vote for Britain to leave Europe (Versi 2016). The students as youth and community workers have experienced this assertive English nationalism *After Empire* (Gilroy 2004), within the context of different communities being impelled to live with different racialised realities in an increasingly socio-economically and spatially divided society. This re-racialisation is most immediately experienced in their daily lives, as a result of the intense surveillance that differentially impacts on diasporian groups and their location, spatially, socio-economically, generationally and subjectively. At a time when commentators speak of a post-racial politics, the students' caseloads in the recent past indicate the historical continuity of racially inflected class-based structural constraints on black and Muslim communities. The latter's collective profile includes the highest levels of unemployment and over-representation in low skilled employment, over-representation in prisons, over-representation in poor housing, high levels of poor health and lowest levels of social mobility (Equality and Human Rights Commission 2016; Garner and Bhattacharyya 2011).

> Minakshi: Austerity has massively increased divisions in this city....poor people, white people as well minority communities have been hit the hardest.

> Marcella: For minorities in Britain, my mother says life has got better because there's not the overt racism of the 70s. But if you look at racism working through poverty and extreme hostility from people in power it's really gone backward. Perhaps, it's also more complex now because, white poor people are also really despised.

As suggested above, a majority of the students tended not to refer to a materialist understanding of racism when discussing their current experience of university life. Rather, in their social biographies, it was invoked in remembering experiences that were marked by various forms of symbolic violence, social exclusion, misrecognition as well as gender specific forms of racialisation in their earlier schooling. The interviewees' narratives were particularly of interest as they had experience of a wide range of educational institutions: primary school; comprehensive and grammar secondary schools; further education; a Russell Group university; and, a children's home. Most of the students' narratives illustrated the pervasiveness of racial exclusion that helped shape their childhood and young adult lives across educational sites. Hence, there is no simple shift from colonial-based understandings of raced subjects and racism to late modern ethnicity within post-colonial conditions. However, the latter might be interpreted as forming a primary focus of many of their narratives about their current pedagogical experience.

> Joanne: Well the teaching here reflects that we're born here, doesn't it? It would be different if it were our parents' generation. Lots of things have changed and got more complex. Like all the talk about identity, our generation, we've grown up with identity politics.

> Iftikhar: And then the crisis about British identity. It's very important especially for young people, young white people I work with. They're looking for an identity too, maybe the most, and they've got all the issues, like unemployment, homelessness, a lot of things against them....with no-one listening to them, except the far right.

Conclusion

We began this research with a focus on exploring current experiences of the anti-racist university. Against the pessimistic trend in the literature concerning the promotion of racial equality and engendering institutional change in H.E. in the post Race Relations (Amendment) Act (2000) period, the students in this qualitative study spoke of their university experience as the place where an emerging pedagogical self was imagined and lived out as part of a wider process, where subjectivities, identifications and identities became reconfigured. The students tended to work with decentred notions of race, ethnicity and racism. As late modern subjects, they spoke of the pedagogical opportunities offered by the youth and community course to reflect on multiple identities and subjectivities that enabled them to re-read their earlier educational biographies – reflecting on the interconnections between their past, present and future. The wide range of theories of racisms and ethnicities to which they were introduced, together with their own cultural capital that the former enabled them to recognise, has helped them to produce sophisticated, nuanced narratives about contemporary racialisation, grounded in their professional biographies as youth and community workers. Most significantly, they are writing their own narratives about their futures.

Disclosure statement

No potential conflict of interest was reported by the authors.

References

Appadurai, Arjun. 1991. "Global Ethnoscapes: Notes and Queries for a Transnational Anthropology." In *Recapturing Anthropology: Working in the Present*, edited by Richard G. Fox, 191–210. Santa Fe, CA: School of American Research.

Archer, Margaret S. 2007. *Making Our Way Through the World*. Cambridge: Cambridge University Press.

Bagguley, Paul, and Yasmin Hussain. 2014. "Negotiating Mobility: South Asian Women and Higher Education." *Sociology* IFirst: 1–17.

Basit, Tehmina, N. 2013. "Educational Capital as a Catalyst for Upward Social Mobility Amongst British Asians: A Tree-Generational Analysis." *British Educational Research Journal* 39 (4): 714–732.

Bauman, Zygmunt. 1996. "From Pilgrim to Tourist or a Short History of Identity." In *Questions of Cultural Identity*, edited by Stuart Hall and Paul du Gay, 18–36. London: Sage.

Bauman, Zygmunt. 2001. *Liquid Modernity*. Cambridge: Polity Press.

Benjamin, Jessica. 1998. *Shadow of the Other: Intersubjectivity and Gender in Psychoanalysis*. New York: Routledge.

Bernstein, Basil. 2000. *Pedagogy, Symbolic Control and Identity: Theory, Research, Critique*. 2nd ed. Lanham: Rowman and Littlefield Publications.

Bhopal, Kalwant, and Patrick Danaher. 2013. *Identity and Pedagogy in Higher Education*. London: Bloomsbury.

Bird, J. John, A. Sheibani Azar and D. Diane Francombe 1992. *Ethnic Monitoring and Admissions to Higher Education*. Bristol: Employment Department/Bristol Polytechnic.

Bourdieu, Pierre 1997. "The Forms of Capital." In *Education, Culture, Economy, Society*, edited by A. H. Halsey, Hugh Lauder, Phillip Brown, and Amy Stuart Wells, 46–58. Oxford: Oxford University Press.

Brah, Avtar, J. Hickman Mary, and Mairtin Mac an Ghaill, eds. 1999. *Thinking Identities: Ethnicity, Racism and Culture*. Basingstoke: Macmillan Press.

Braun, Virginia, and Victoria Clarke. 2006. "Using Thematic Analysis in Psychology." *Qualitative Research in Psychology* 3 (2): 77–101.

Burke, Penny Jane, Gill Crozier, Barbara Read, Julia Hall, Jo Peat, and Becky Francis. 2012. *Formations of Gender and Higher Education Pedagogies*. Gap, Final Report, the Higher Education Academy, National Fellowship Scheme. London: University of Roehampton.

Butler, Judith. 2004. *Undoing Gender*. London: Routledge.

Crouch, Mira, and Heather McKenzie. 2006. "The Logic of Small Samples in Interview-Based Qualitative Research." *Social Science Information* 45 (4): 483–499.

Crozier, Gill, and Jane Davies. 2006. "Family Matters: A Discussion of the Bangladeshi and Pakistani Extended Family and Community in Supporting the Children's Education." *The Sociological Review* 54 (4): 678–695.

Equality and Human Rights Commission. 2016. Healing a Divided Community: The Need for a Comprehensive Race Equality Strategy. Equality and Human Rights Commission. Accessed 18 August 2016. https://www.equalityhumanrights.com/sites/default/files/healing_a_divided_britain_-_the_need_for_a_comprehensive_race_equality_strategy.pdf

Fadeeva, Zinaida, and Yoko Mochizuki. 2010. "Higher Education for Today and Tomorrow: University Appraisal for Diversity, Innovation and Change towards Sustainable Development." *Sustainability Science* 5: 249–256.

Foucault, Michel. 1978. *An Introduction. Volume 1 of the History of Sexuality*. New York: Random House.

Foucault, Michel. 1988. "Technologies of the Self." In *Technologies of the Self: A Seminar with Michel Foucault*, edited by Luther H. Martin, Huck Gutman and Patrick H. Hutton, 16–50. London: Tavistock.

Freire, Paulo. 1972. *Pedagogy of the Oppressed*. Harmondsworth: Penguin.

Garner, Steve, and Gargi Bhattacharyya. 2011. *Poverty, Ethnicity and Place*. York: Joseph Rowntree Foundation.

Gilroy, Paul. 2004. *After Empire: Melancholia or Convivial Culture*. London: Routledge.

Hall, Stuart. 1992. "The Question of Cultural Identity." In *Modernity and Its Futures*, edited by Stuart Hall, David Held and Tony McGrew, 273–327. Cambridge: Polity.

Haywood, Chris, and Mairtin Mac an Ghaill. 2013. *Education and Masculinities: Social, Cultural and Global Transformations*. London: Routledge.

Lave, Jean, and Etienne Wenger. 1991. *Situated Learning: Legitimate Peripheral Participation*. Cambridge: Cambridge University Press.

Law, Ian. 1996. *Racism, Ethnicity and Social Policy*. London: Prentice Hall.

Lucey, Helen, Jane Melody and Valerie Walkerdine. 2006. Uneasy Hybrids: Psychosocial Aspects of Becoming Educationally Successful for Working Class Young Women. In *The Routledge Falmer Reader in Gender and Education*, edited by Madeleine Arnot, and Mairtin Mac an Ghaill, 238–251. London: Routledge.

Mac an Ghaill, Mairtin. 1999. *Contemporary Racisms and Ethnicities: Social and Cultural Transformations*. Buckingham: Open University Press.

Mac an Ghaill, Mairtin, and Chris Haywood. 2014. "British Born Pakistani and Bangladeshi Young Men: Exploring Unstable Concepts of Muslim, Islamophobia and Racialization." *Critical Sociology* 41 (1): 97–114.

Martin, Luther ,H., and Gutman, H. Huck, and Hutton Patrick, eds. 1988. *Technologies of the Self: A Seminar with Michel Foucault*. London: Tavistock.

McLean, Monica. 2006. *Pedagogy and the University: Critical Theory and Practice*. London: The Continuum International Publishing Group.

Miah, Shamim. 2015. *Muslims, Schooling and the Question of Self-Segregation*. Basingstoke: Palgravemacmillan.

Mirza, Heidi, S. 2015. "Decolonizing Higher Education: Black Feminism and the Intersectionality of Race and Gender." *Journal of Feminist Scholarship* 7 (8): 1–12.

Modood, Tariq. 2012. "Capitals, Ethnicity and Higher Education." In *Social Inclusion and Higher Education* edited by Tehmina N. Basit, and Sally Tomlinson. Bristol: Policy.

Nixon, Sean. 1997. "Exhibiting Masculinity." In *Representation: Cultural Representations and Signifying Practices*, edited by Stuart Hall, 291–336. London: Sage, The Open University Press.

Race Relations (Amendment) Act. 2000. *Home Office*. London: UK Government.

Said, Edward. 1993. *Culture and Imperialism*. London: Vintage.

Sallah, Momomdou, and Carlton Howson. 2007. *Working with Black Young People*. Russell House: Lyme Regis.

Ussher, J. M, M Sandoval, J Perz, W. K. T. Wong, and P. Butow 2013. The Gendered Construction and Experience of Difficulties and Rewards in Cancer Care, Qualitative Health Research. http://qhr.sagepub.com/content/early/2013/04/03/1049732313484197

Versi, Miqdaad. 2016. "Brexit Has given Voice to Racism – And Too Many Are Complicit." *The Guardian*, 27 June 2016. https://www.theguardian.com/commentisfree/2016/jun/27/brexit-racism-eu-referendum-racist-incidents-politicians-media

Affirmative action in Brazil and building an anti-racist university

Joaze Bernardino-Costa and Ana Elisa De Carli Blackman

ABSTRACT

This article highlights the black movement's centrality to building anti-racist universities in Brazil. It examines the questioning of the racial democracy myth within Brazilian universities as well as in the Brazilian mainstream media since the beginning of the new millennium. This debate was referred to the Supreme Court, which affirmed the constitutionality of racially oriented affirmative action measures. After the Supreme Court pronouncement, the Federal Government approved Law 12.711/2012 that instituted affirmative action within every federal university in the country. Even if it seems to be a victory in the anti-racist struggle, in this new law the racial dimension is a category tagging along behind that of social class. Considering that, the article concludes by calling for the need to redeem the original meaning of the debate, namely the anti-racist struggle.

Introduction

If Brazil was imagined to be a race-less nation in much of the twentieth century, we cannot say the same in the twenty-first century. Over the past century the image of a racial democracy, where race was irrelevant in defining life's chances and opportunities, has spread to serve as an example to the world. The black movement and black intellectuals protested, claiming that this was more of a myth than a description of reality. Although we find discordant accounts (Gonzáles 1984; Moura 1977; Nascimento 1978, 2003), these accounts were rejected. In the last decade of the twentieth century this myth was demolished (Guimarães 1999; Henriques 2001). Instead of this myth Brazil emerged a country where race is a fundamental dimension in determining people's life chances and opportunities.

What makes Brazil a unique and intriguing case is the way racism functions and, therefore, race as an important dimension of social life. Talking about race is avoided, even whilst race forms part of the code of conduct of all Brazilians because racial segregation and preferences abound. What we have is a profound racism without racists (Bonilla-Silva 2014).

This racism without racists is a direct product of the myth of racial democracy, formulated in the 1930s, by Gilberto Freyre. In 'The Master and the Slaves', Freyre (1986) describes Portuguese colonisation as benign, marked by fraternising between blacks and whites. Unlike other racial formations elsewhere, in Brazil racial barriers would not have arisen, resulting in a country in which its main and much feted symbol would be the person of mixed race. Moreover, within the myth of the racial democracy account inequalities between whites and blacks could not be attributed to race, but rather to class discrimination and the effects of past enslavement.

Thus, racism in Brazil has rarely been described explicitly. Throughout the twentieth century, for example, compared to the United States or South Africa, we can rarely find a state declaration that race discrimination existed (Goldberg 2002). However, the effects of race and racism can be grasped indirectly through statistics. Several recent studies have shown the effects of racial discrimination in education. For example, even after almost a decade of implementation of affirmative action in Brazilian universities, 21.6% of whites and 8.3% of blacks, between 18 and 24 years old, were in higher education in Brazil in 2009, although the country is composed of 50.7% blacks and 47.7% whites (IPEA 2012).[1] If racial inequality at the end of the first decade of the new millennium was still glaring, fifteen years ago, when affirmative action was adopted in Brazilian universities, it was even more so. In 1999, 11% of white youths aged between 18–24 were in universities compared with only 2% of black young people (Henriques 2001).

When these figures began to surface at the end of the last millennium, a great movement was triggered in favour of affirmative action for the black population in different universities in the country. However, it is important to note that the statistical discovery of racial inequalities and the struggle of the black movement for equality in the education system were present in Brazilian society from the middle of the last century (Bernardino-Costa and Rosa 2013). Since Carlos Hasenbalg's (1979) work on the social mobility of blacks and whites in Brazil, racial discrimination processes have come to light in the educational system and in the Brazilian labour market. This pioneering study, contradicted the notion that racial inequalities were only a product of past slavery and demonstrated that racism is present in Brazilian social relations. Even earlier than this, dating from the 1940s, there are documents in which the black movement demanded affirmative action in education (Nascimento 2003).

Both the message of the black movement and the studies undertaken came to focus increasingly on the racial aspect of social inequalities in the country, because social class has proved insufficient in understanding social inequality in Brazil. Along with this how inequalities based on race are dealt with has also changed. The anti-racist struggle, then, has come to be expressed more in terms of specific public policies directed at the black Brazilian population, rather than in terms of policies of a universal nature. Since universal policies have not managed to combat inequalities, more specific public policies have been structured to address the realities of black people, women and disabled people, among others.

However, in the late 1990s, there was a combination of actors that led to the historic demands of the black movement gaining a favourable reaction in Brazilian politics, resulting in widespread adoption of affirmative action in the policies of Brazilian public universities (Feres Júnior and Toste Daflon 2014). In 2012, a decade after the first university[2] adopted affirmative action policies, and the ensuing intensive public debate in the country involving the mainstream media, the Supreme Court unanimously approved the constitutionality of affirmative action for blacks in the country. Soon after the federal Law 12.711/2012 was approved, students from public schools, poor students, and black and indigenous students in the federal universities of Brazil had guaranteed access.

In the subsequent parts of this article we first explain how the country abandoned the belief in the myth of racial democracy and began to adopt affirmative action policies in public universities. Second, we address the public debate in Brazilian society about affirmative action, which peaked with the judgment by the Supreme Court in the Claims of Non-Compliance with a Fundamental Precept (ADPF n°. 186),[3] presented by the right-wing party, the Democrats. Third, we comment on the adoption of Law 12.711/2012 (the so-called Quota Law) and its impact on the anti-racism struggle within universities. Finally, the article concludes by drawing attention to the fact that the anti-racist struggle in universities does not end with the guarantee of access to undergraduate courses, but has other challenges.

From the myth of racial democracy to the adoption of affirmative action policies

As has already been stated, Brazilian race relations are marked by the idea of 'the non-existence of strong racial lines' (Guimarães 1999) in the distinction between black people and white people, giving the impression that the country is a racial paradise because of the absence of prejudice and discrimination based on race. This is a perennial vision originating from studies relative to slavery in Brazil, which is supposed to have been 'milder' and 'more human' when compared with slavery in other contexts.

Subsequently, the notion of a racial paradise has come to refer to the overcoming of the trauma of slavery on the part of Brazilian society, which went on to incorporate black people in a positive way with the adoption of a mixed-race or syncretic national culture (Guimarães 1999).

Through the decades, racial democracy, especially from the political point of view, has come to be questioned by the black movement. In addition to this, it is evident that there is a great distance between black people and white people from the economic and social point of view, and therefore a section of Brazilian society has come to denounce the myth of racial democracy.

This whole context has led to polarisation between the two interpretations with respect to Brazilian racial reality. On one hand, there are interpretations along the lines of racial democracy in which the founding myth of the uniqueness of the constitution of Brazilian society as a nation continues to guide social relations between white people and black people (Guimarães 1999). On the other hand, there are interpretations which denounce the myth of racial democracy as a strategy of domination, because it has not incorporated the black population socially, economically and politically.

A landmark in the antiracist struggle was the march held on November 20, 1995, the day of black consciousness, in the capital, Brasilia, which was decisive for the country's change of stance as a racial democracy. At the end of that march, attended by about 30,000 people, the 'Overcoming Racism and Racial Inequality Programme' was delivered to the President of the Republic. The proposals made in this programme explicitly mentioned affirmative action for black students in Brazilian universities.

As a direct impact of this march, President Fernando Henrique Cardoso (in office from 1995 to 2002), in addition to other measures, promoted the international seminar on 'Multiculturalism and Racism: The role of affirmative action in contemporary democratic states', in July 1996. In this seminar, for the first time in Brazil, the head of state admitted the existence of racial discrimination:

> Here we have discrimination, here prejudice (…) Discrimination has established itself as something that is being repeated constantly, which is reproducing itself. And it will not do for hypocrites to say, "No, this is not our way". No, our way is simply wrong. Discrimination is repeated continually. This has to be unmasked. It has to be really counter-attacked, not only verbally but also in terms of mechanisms and processes that can lead to a transformation toward a more democratic relationship between races, between social groups, between classes. All this has to be done (Cardoso 1998, 7–9).

Thus, the end of the 1990s was marked by numerous measures indicating the recognition of racism in Brazilian society by the Federal Government. The adoption of affirmative action policies would gain a new impetus with the preparatory discussions for the 'Third World Conference against Racism, Racial Discrimination, Xenophobia and Related Intolerance', held by the United Nations in 2001. In 2000 and 2001, several regional conferences and debates, in which black activists took a prominent part, were held in the country. These discussions assisted in the preparation of the Brazilian document that was sent to this UN conference held in Durban, South Africa, from August 30 to September 7, 2001.

In this conference's 'The Declaration and the Programme of Action', the signatory countries, including Brazil, undertook to develop policies to combat racial inequalities, including affirmative action policies, and to recognise race as an explanation for the profound social inequalities existing in Brazil. The Plan of Action stated in paragraphs 99 and 100:

> Paragraph 99 - Recognizes that combating racism, racial discrimination, xenophobia and related intolerance is a primary responsibility of the States. It therefore encourages States to develop and elaborate national action plans to promote diversity, equality, fairness, social justice, equal opportunity and participation for all. Through, among other things, affirmative or positive actions and strategies (…) Paragraph 100 - Urges the States to establish, based on statistical information, national programs, including programs of affirmative action or positive action measures to promote access for groups of individuals who are or may become victims of racial discrimination in basic social services, including basic education, primary health care and adequate housing (Moura and Barreto 2002, 131).

In 2001, the Ministry of Agrarian Development, the Supreme Federal Court (Constitutional Court), the Ministry of Justice and the Ministry of Culture established 20% quotas for blacks in the hiring of temporary workers. In addition, they established priority contracts with companies providing services which had adopted affirmative action programmes for black people (Htun 2004). The UN conference led to the implementation of legal support for and strengthening of the pro-affirmative action offensive, especially in public universities.

Reflecting the UN conference on November 9, 2001, the Rio de Janeiro State Legislative Assembly approved State Law 3708, reserving 40% of the places in the State University of Rio de Janeiro (UERJ) and the State University of North Fluminense (UENF) for 'black and brown' students. The reservation of places for 'black and brown' students was preceded by Law 3524 of December 28, 2000, which reserved 50% of the places in the universities mentioned above for students coming from public schools. In the first entrance exam under this system, 90% of the places were reserved for the quota system (Daflon, Feres Júnior, and Campos 2013).

Unlike the Brazilian university system, public elementary and high schools are usually 'lower quality' institutions, while private elementary and high school are 'higher quality' ones. The inequality system in the country works like that: rich and middle class families enrol their children in private elementary and high school institutions, after that these students (usually white students) go to public and high quality universities. On the other hand, working-class families enrol their children in public elementary and high school institutions, after that these students (usually black students) don't get places at public universities and go to private colleges and universities, which are not as good as public ones. Here is one of the keys to understanding racial and social inequality in Brazil.

After a barrage of criticism and public debate on the quota system's and high percentages, state laws were changed. Law 4151 of 4th September 2003 set 20% reservation for students from public schools, 20% for black students at UERJ and UENF.

In 2002, the Bahia State Legislative Assembly also passed a law instituting 40% of reserved places for black students and those originating from state schools in undergraduate and graduate courses in the State University of Bahia (UNEB).

The adoption of affirmative action in the case of state universities above had come from outside the decisions of the academic community through the respective Legislative Assemblies. However, on June 6, 2003, the University of Brasilia's Administrative Board approved the Plan for Ethnic-Racial Aims and Integration, which provided for the reservation of 20% of the undergraduate course vacancies of that university for black students. What motivated this decision by the University of Brasilia was the failure of a black Doctoral student in Social Anthropology at this University in 1998. This event sparked an intense debate at the University of Brasilia, which culminated in the adoption of the affirmative action policy. The new departure in the 'Plan for Ethnic-Racial Aims and Integration' at the University of Brasilia was not simply the fact of the decision being made by the academic community, but the fact of the affirmative action policy being entirely based on race, despite the socioeconomic status of students and the type of school attended.

After the pioneering decisions of state universities mentioned above and of the prestigious University of Brasilia, several universities fought long battles in their deliberative councils. Similarly, some state legislatures also had discussions on the adoption of affirmative action policies. The process has been characterised by a number of contradictions, as well as opposition from traditional segments of society, but these have not restricted the debate nor the legitimacy of such measures.

In May 2012, before the adoption of a federal law on the subject (Law 12,711/2012), of the ninety-six state and federal public universities in the country, seventy adopted some form of affirmative action (Daflon, Feres Júnior, and Campos 2013).

As the decision to adopt an affirmative action programme went out from state legislatures and the deliberative councils of each of the federal universities, these 70 institutions each adopted their own affirmative action programmes (INCT – Instituto de Inclusão no Ensino Superior e na Pesquisa 2013).

Resulting from the autonomy of the universities and the autonomy of the states in adopting their affirmative action programmes, higher education institutions have implemented three types of affirmative action: quotas, bonuses or awards and additional place creation[4] for a given population group. Of these three types of affirmative action, the vast majority adopted the quota system.

Another factor that varied greatly due to the autonomy of the universities and states was the beneficiary group. Although the struggle for affirmative action was triggered by the black movement and anti-racist intellectuals, the racial dimension was subordinated to the question of class (Santos 2015). In many universities, the affirmative action programme gave priority to students from public schools and among these a certain percentage of black students. In others, the programme is only intended for students from public schools, without any mention of race. Of all the seventy universities in 2012, who had some affirmative action policy, sixty had programmes for public school students, forty-one for black students and thirty-six for indigenous students. Here are some examples of how the system operated:

- The University of Brasilia (UNB) adopted its affirmative action programme in 2003, reserving 20% of places for black students regardless of family income and of the type of school where the students had completed their high-school course. In addition, it created ten additional places according to the demand in any undergraduate programme for indigenous students.
- The Federal University of Bahia (UFBA) approved its affirmative action programme in 2004, allowing 36.55% for black and brown public school students, 6.45% for public school students of any ethnicity or race and 2% of vacancies for students who declared themselves to be descendants of indigenous people.
- The Federal University of Santa Catarina (UFSC) approved its affirmative action programme in 2007, reserving 20% of places for students who have had all elementary and high school education in public institutions and 10% for black students who have completed elementary and high-school education in public institutions.
- In 2010, the Federal University of Rio de Janeiro, the largest federal university in the country, approved a resolution to reserve 30% of places for state school students with per capita income up to one (1) minimum wage, without giving preference to any racial group.

From the few examples given above, affirmative action programmes in the country appear to be heterogeneous (INCT – Instituto de Inclusão no Ensino Superior e na Pesquisa 2013). This diversity would only be regulated in 2012, when the federal government passed the Law 12,711.[5]

The change in the form of access to prestigious universities in Brazil caused a profound debate in Brazilian society. The debate continued practically every day not only within university classrooms but also in the press. The culmination of this debate was the vote by the Supreme Constitutional Court for the affirmative action programme at the University of Brasilia, which was based exclusively on race.

The public debate: is affirmative action constitutional?

The decision of the state universities of Rio de Janeiro in 2001, and that of the University of Brasilia in 2003, brought the debate on access to higher education on to the agenda. The country's main newspapers began to fill their pages with reports on Brazilian universities (Feres Junior, Campos and, Daflon 2011).

On the one hand, the main arguments of those opposed to affirmative action policies stated that (a) these policies violated the principle of merit; (b) affirmative action reinforced the prejudice against blacks; (c) the problem in Brazil was with elementary and high school education; (d) we do not know how to define who is black in a country of mixed races and (e) the absence of blacks in universities and in prestigious positions in Brazilian society is not because of race, but because of class (poverty).

On the other hand, those favourable to affirmative action policies argued that (a) the discussion on their merits cannot be reduced to the discussion of the proof of access to a university, but must take into account the trajectory of the students; (b) the affirmative action policies are necessary in order to have blacks in prominent positions, play the part of role models in Brazilian society and thus help to reverse the prejudices that affect the black population; (c) as well as long-term investments in basic education, we must invest in short-term policies focused on the black population; (d) it is undeniable that there is a high degree of racial mixture in the country, but we know who is black in Brazil and (e) class differences explain in part the social inequalities in Brazil, but race is an important factor of disadvantage for the black population, and should therefore be considered in the development of public policies (Carvalho 2006; Gomes 2001; Guimarães 1999; Munanga 2003; Paixão 2006; and Santos 2003).

These arguments – both for and against – appeared again on two occasions in the Brazilian public arena. On 30 May 2006, a group of university professors and members of political parties gave the President of the National Congress a document entitled 'Public Letter to Congress - Everyone has rights in the Democratic Republic'. In this manifesto, contrary to affirmative action policies, the main argument was that the approval of the Quota Law[6] would create a racialised state and that history had shown the adverse effects of this type of policy. Therefore, the state should not support laws that differentiated between its citizens, because all were equal before the law. In response, university professors, intellectuals, activists and militants in the black movement supportive of affirmative action policies sent 'The Manifesto in Favour of Quotas Act and the Statute of Racial Equality'[7] to the President of the National Congress on 03 July of that year. This manifesto criticised the signatories of the first manifesto, claiming that the defence of formal equality had been the prevailing attitude after the abolition of slavery in the country in 1888. At that time the Brazilian state proclaimed the equality of all before the law, but for nearly half a century adopted incentives, policies and support for European immigration. 'The Manifesto in Favour of the Quotas Act and the Racial Equality Statute' argued that the Brazilian State was and still is racially biased against the black population. It was therefore necessary to adopt affirmative action policies in order to achieve true racial equality in the country.[8]

Another public staging of this debate would take place two years later. On 30 April 2008, again the group opposing the adoption of affirmative action policies gave the President of the Supreme Court 'The Manifesto One-Hundred and Thirteen Anti-Racist Citizens against the Race Laws'.[9] In this Manifesto the arguments of the 'Public Letter to Congress - Everyone has rights in the Democratic Republic' were reiterated, and alerted the President of the Supreme Court of the risk of creating racial laws. The manifesto equated affirmative action policies already underway in the country at that time in many universities, with the creation of racial distinctions in the United States by Jim Crow, in South Africa by apartheid, in Germany by Nazism. The *Manifesto* concluded with the warning that affirmative action, if approved, could create insurmountable boundaries between whites and blacks in Brazil.

Again there was a response from those in favour of affirmative action. On a symbolic day, 13 May 2008, the day which celebrates the abolition of slavery in Brazil, the document '120 Years of Struggle for Racial Equality in Brazil: A Manifesto in Defence of Justice and Constitutional Quotas', was submitted to the President of the Supreme Court. This document responded to the manifesto against Quotas. It argued that affirmative action now underway in the country had already offered in five years more opportunities for black students than in the whole of the twentieth century. In addition, throughout this time, racial conflicts were not witnessed in Brazil, such as those who were against such policies had claimed. It also argued that the adoption of affirmative action was consistent with international agreements to overcome racism, of which the country is a signatory.

The culmination of this public debate took place on 25th and 26th April 2012 when the Supreme Court ruled on the Claims of Non-Compliance with a Fundamental Precept (ADPF nº. 186), filed in July 2009 by the Democrats, one of the most conservative political parties of the country's right wing. In a 612-page document with annotations attached, some academics pointed out that the manifestos against the country's affirmative action policy, the ADPF nº. 186 called for the declaration that the affirmative action programme at the University of Brasilia (UNB) was unconstitutional and consequently, the extension of the declaration to all similar programmes. It is clear that the main argument of the ADPF was not against affirmative action as a form of public policy, but against the use of racially oriented affirmative action. In fact the ADPF nº 186 argued for the adoption of affirmative action based on social class.

The ADPF nº 186 did not present any new arguments beyond what had already been presented so far in the public debate. Basically it presented the following arguments: the criticism of the concept of race; defence of the benign nature of slavery in Brazil and the non-culpability of the white masters; that the exclusion of blacks from prestigious positions was not the result of racial discrimination, but of poverty, and finally, a comparison between Brazil and the United States.

The criticism of the concept of race that appears in this document is based on mistaken assumptions. It argues that the proponents of affirmative action were operating with a sense of biological races but races do not exist. What ADPF does not clarify is that the concept of biological races was never used in the debate on affirmative action. As is widely known, race is a social construct which defines individual life chances and opportunities (Osório 2003; Soares 2008; Theodoro 2008).

The last argument, regarding the risk that Brazil would resemble the United States from the time of Jim Crow as a result of affirmative action policies also makes absolutely no sense. There is no way in which we could equate Brazilian policies aimed at affirmative action with the dehumanising policies and exclusions seen in segregationist US policy.

After three days of public hearings (*amicus curiae*), on 25th and 26th April 2012, the ministers of the Supreme Court unanimously accepted the vote of the rapporteur, Minister Ricardo Lewandowski, rejecting ADPF nº. 186 and ensuring the constitutionality of racially oriented affirmative action as adopted by the University of Brasilia and other universities in the country.

The opinion of Minister Ricardo Lewandowski defended the constitutionality of affirmative action for blacks in Brazil and reaffirmed the commitment of the Constitution of Brazil to the principle of material equality. Thus, to achieve substantive or material equality if it was necessary to use race as a social-historical category, then social institutions could do so (Lewandowski 2013)

Four months after the Supreme Court dismissed the Claims of Non-Compliance with a Fundamental Precept (ADPF nº. 186) filed by the Democrats and consequently declared constitutional the adoption of affirmative action for blacks in Brazil, the National Congress (House of Representatives and Senate) passed Law 12,711/2012 on the 29th of August.

Law 12,711/2012 and the anti-racist struggle in Brazilian universities

As mentioned above, prior to the enactment of Law 12,711/2012 or the Quota Law,[10] as it was named in the public debate, there was a heterogeneous system of affirmative action.

This heterogeneity was due to the fact that each university had autonomously decided on the adoption of its own affirmative action programme. Until 2012 the system adopted was so complex and differentiated that we can say, with great probability of being correct, that each of the thirty-nine federal universities[11] had its own affirmative action programme with variations according to the form of affirmative action adopted (quotas, bonuses or additional spaces created), in beneficiaries (students from public schools, blacks, indigenous people) and in percentages of places reserved for black students, for students from public high school, for indigenous students and so on.

Overall, the new legislation established a system that combined quotas for students from state schools and two types of sub-quotas. First, sub-quotas for students from families with incomes at or below 1.5 times the minimum wage. Second, sub-quotas for black (*preto*), brown (*pardo*), and indigenous (*indígena*) students, in proportion to the number of black, brown and indigenous people in the state in which the university is located. Let us take a look at the text of the law:

> Art. 1 - The federal institutions of higher education linked to the Ministry of Education will reserve, in each selection competition for admission to undergraduate courses, by course and group, at least 50% (fifty per cent) of its places for students who have completed secondary education in state schools.

> In filling places referred to in this article, 50% (fifty per cent) should be reserved for students from families with income equal to or less than 1.5 of the minimum wage (one minimum wage and a half) per capita.

> Art. 3 - In each federal institution of higher education, the places mentioned in art. 1 of this Law shall be filled, by course and group, by self-declared black, brown and indigenous students, in proportion at least equal to that of black, brown and indigenous people in the population of the unit of the Federation where the institution is installed, according to the latest census of the Brazilian Institute of Geography and Statistics. (BRASIL, 2012a)[12]

Thus, for example, a university that offers 8000 places per year, must reserve places for 4000 public school students. Of these 4000 places, at least 2000 should be reserved for students from families with incomes at or below 1.5 of the minimum wage per capita. If the university is located in a state where 56% of the population is black, at least 2240 places must be filled by black students, divided between those with incomes at or below 1.5 times the minimum wage per capita.

The impact of the Quota Law on the anti-racist struggle in Brazilian universities is yet to be thoroughly assessed. Even with the Supreme Court decision in favour of racially oriented affirmative action policies, the scenario was uncertain. Universities in Brazil decided to adopt affirmative action programmes for a period of ten years, after which such policies would be evaluated, and could be maintained, revised or eliminated. Although the Supreme Court decision made affirmative action based on race constitutional, it did not require universities to adopt it. In fact, it did not even oblige universities to adopt an affirmative action programme. Brazilian universities could go back to what they were before the movement for affirmative action and have places dominated by whites.

The struggle for affirmative action in Brazil was a struggle triggered by the few professors and black students already in the universities and by the black social movement trying to make the university entry rules more just and democratic. Although some universities, like the University of Brasilia, have been able to adopt a programme based solely on race, in many other universities black professors strongly committed to combating racism and racial exclusion within universities had to make concessions and changes in affirmative action projects before they could be approved. Thus, in many universities the racial component of the affirmative action policy was not so prominent.

What is certain is that the Quota Law could put programmes strongly committed to the fight against racial exclusion within Brazilian universities in jeopardy, such as the University of Brasilia, because it subordinated the racial dimension to the class dimension (public school and family income). In other words, race had to be second behind class. In many other cases, such as the Federal University of Rio de Janeiro, the law ensured at least some consideration for the racial dimension. Due to the extreme conservatism of its faculty,[13] this university later adopted a law providing a quota of 30% of its places for state school students, taking into account the family income of the student. With the Quota Law, the university had to also reserve places for black students. So for this university, Law 12,711/2012 meant a breakthrough in the fight against racial exclusion.

If an affirmative action policy is racially oriented, or if race tags along behind social class, it makes a big difference in the meaning of the policy. Programmes strictly based on race, such as in the University of Brasilia, were strongly committed to anti-racist struggle. However, affirmative action programmes, such as that approved by Law 12,711/2012 and which apparently[14] predominated in Brazilian universities before that law, are programmes in which the anti-racist struggle is not necessarily foregrounded.

From the point of view of numbers, it appears that Law 12,711/2012 was a victory in anti-racist struggle in Brazilian universities. However, from the point of view of the meaning of the law, it is clear that the race issue that motivated and motivates the discussion on affirmative action, lost centrality and came to be seen as less important than social class. Thus, almost fifteen years after the adoption of affirmative action in the first Brazilian university, the task is to get back to the original meaning of the anti-racist struggle.

Conclusion: new challenges in the anti-racist struggle in Brazilian universities

Despite our sympathy with the idea that the university should not be an area for the reproduction of the economic elite of the country – hence our support also for policies based on social class – it is necessary to restore the centrality of race in the debate on affirmative action in Brazil. As mentioned above, the struggle for the adoption of affirmative action policies was linked to black activism since the 1990s. For this reason, the first programmes were based on race. The Law that came to regulate inequality in universities, on the other hand, considers race as a sub-quota category (Santos 2015). What is at stake again, at the present time, is the dilemma between class and race in Brazilian society. It is essential to return to the initial demand of the anti-racist struggle in favour of affirmative action.

When the first universities adopted their affirmative action programmes at the beginning of the millennium, they were linked to the fight against racism in Brazilian universities. The main argument in the debate was the absence of black students. With the adoption of the Quota Law this argument became secondary, class inequalities in Brazilian society becoming the main reason for the adoption of this law. It was believed that in resolving the economic issues the racial issue will be resolved automatically. However, numerous studies over the past century have not corroborated this conclusion (Hasenbalg 1979; Henriques

2001). Despite the country's economic development and general improvement at all socio-economic levels, without the adoption of racially oriented policies, racial inequalities were maintained or even increased (Henriques 2001).

Upon entering the university via the Quota Law, it is essential to recognise that a black student is in the university not because of class, but because of the anti-racist struggle. For the new generations of black students to contribute to the fight against racism in Brazilian universities and Brazilian society in general, it is critical that these students recognise themselves as black and not as second-class whites. However, for this to be done it is essential to recover the sense of the anti-racism struggle that democratised access to universities.

The challenge is to rewrite Brazilian history and that of the African diaspora, recovering the importance and contribution of black people in the formation of the modern world. In this case, it is urgent to demolish the epistemological racism that prevails in Brazilian universities. This epistemological racism assumes that only one racial group is capable of producing knowledge and that the presence of another group systematically excluded from the production of knowledge, will corrupt Brazilian academic achievements.

The struggle against epistemological racism will promote the democratisation and pluralization of knowledge. Once this is achieved, we may realise that which was also the desire of Frantz Fanon (2008), that we make sense of the world from our own positionality and from our own racial condition.

Notes

1. The black population (negro) is composed of those who self-identify as black (*preto*) and brown (*pardo*) in research conducted by the Brazilian Institute of Geography and Statistics (IBGE). In the last Census (2010) the black population totaled 50.7% of Brazilians – 7.6% of *preto* and 43.1% self-identified as *pardo*.
2. As we will show in this article, State University of Rio de Janeiro (UERJ) and State University of North Fluminense (UENF) were the first universities that adopted affirmative action in 2001.
3. *Arguição de Descumprimento de Preceito Fundamental* (ADPF) or the Claims of Non-Compliance with a Fundamental Precept is a constitutional control device which can be presented by any citizen or association that believe their rights were violated.
4. The creation of additional places was applied more to the country's indigenous population. For example, the Plan for Ethnic-Racial Aims and Integration at the University of Brasilia (UNB) created 10 additional places in each selection process (two per year), according to the demand of indigenous groups.
5. Note that Law 12,711/2012 applies only to institutions within a federal system, that is, to 59 institutions, in addition to Federal and Technical Education Centres (CEFETs), which we are not considering in this article. Another 37 higher education institutions belong to the state and are not subject to this Law.
6. The Quota Law (Law 12,711/2012), approved in 2012, had been discussed since 1999, because of that, *The Public Letter to Congress – Everyone has Rights in the Democratic Republic* mentioned it in 2006.
7. The Statute of Racial Equality is a set of legal guidelines for the adoption of targeted policies for the Brazilian population. Among the specific policy for the black population, the Statute provides for laws on education, land, health, the labour market. The Statute of Racial Equality was presented to Congress in 2000. However, it was only approved in July 2010 (Law 12,288). For details, see: Silva 2012.
8. Cf. http://www1.folha.uol.com.br/folha/educacao/ult305u18773.shtml
9. Cf. http://www1.folha.uol.com.br/fsp/cotidian/ff1405200807.htm

10. Remember that this law applies to Federal Universities and Federal Technological Education Centres; it does not apply to municipal, state and private higher education institutions. In this paper, we refer only to the Federal Universities.
11. As mentioned earlier, there are 59 federal universities in the country. Thus, 2/3 of them had adopted some kind of affirmative action policies when Law 12,711/2012 was approved.
12. Subsequently, the law was regulated by another decree which had to do specifically with the forms of admission to such federal institutions of higher education. For more details see http://www.planalto.gov.br/ccivil_03/_ato2011-2014/2012/Decreto/D7824.htm
13. Some of the professors who prepared the manifestos against the policies of affirmative action, which were delivered to the Presidents of the National Congress in 2006, and to the Supreme Federal Tribunal in 2008, belonged to this university.
14. A detailed survey of the affirmative action programmes before the publication of law 12,711/2012 has yet to be done.

Disclosure statement

No potential conflict of interest was reported by the authors.

References

Bernardino-Costa, Joaze, and Fernando Rosa. 2013. "Appraising Affirmative Action in Brazil". In *Affirmative Action, Ethnicity, and Conflict,* edited by Edmund Terence Gomez, and Ralph Premdas, 183–203. London and New York: Routledge.

Bonilla-Silva, Eduardo. 2014. *Racism Without Racists: Color-blind Racism and the Persistence of the Racial Inequalities in America.* Lanham: Rowman & Littlefield Publishers.

BRASIL (2012a). *Lei n. 12,711, de 29 de agosto de 2012. Dispõe sobre o ingresso nas universidades federais e nas instituições federais de ensino técnico de nível médio e dá outras providências.* http://www.planalto.gov.br/ccivil_03/_ato2011-2014/2012/lei/l12711.htm

Cardoso, Fernando Henrique. 1998. *Construindo a democracia Racial: atos e palavras do Presidente Fernando Henrique Cardoso (1995–1998). Brasília, Presidência da República.* http://www.biblioteca.presidencia.gov.br/ex-presidentes/fernando-henrique-cardoso/publicacoes-1/construindo-a-democracia-racial

Carvalho, José Jorge. 2006. *Inclusão Étnica e Racial no Brasil: A Questão das Cotas no Ensino Superior.* São Paulo: Attar.

Daflon, Verônica Toste, João Feres Júnior, and Luiz Augusto Campos. 2013. "Ações afirmativas raciais no ensino superior público brasileiro: um panorama analítico". *Cadernos de Pesquisa* 43 (148): 302–327. http://www.scielo.br/pdf/cp/v43n148/15.pdf

Fanon, Frantz. 2008. *Pele Negra, Máscaras Brancas.* Salvador: EdUFBA.

Feres Júnior, João, Luiz Augusto Campos, and Verônica Toste Daflon. 2011. "Fora de Quadro: A Ação Afirmativa nas Páginas d'*O Globo*". *Contemporânea – Revista de Sociologia da UFSCar* 1 (2): 61–83. http://www.contemporanea.ufscar.br/index.php/contemporanea/article/view/37

Feres Júnior, João, and Verônica Toste Daflon. 2014. "Políticas de Igualdade Racial no Ensino Superior". *Cadernos do Desenvolvimento Fluminense* 5 (1): 31–44. http://www.e-publicacoes.uerj.br/index.php/cdf/article/view/14229/10769

Freyre, Gilberto. 1986. *The Masters and the Slaves.* Berkeley: University of California Press.

Goldberg, Theo David. 2002. *The Racial State.* Massachusetts/Oxford: Blackwell Publishers.

Gomes, Joaquim Barbosa. 2001. *Ação Afirmativa e o Princípio Constitucional da Igualdade.* Rio de Janeiro: Renovar.

Gonzáles, Lélia. 1984. "Racismo e Sexismo na Cultura Brasileira." *Revista Ciências Sociais Hoje* 2 (1): 223–244.

Guimarães, Antônio Sérgio Alfredo.1999. *Racismo e Anti-racismo no Brasil.* São Paulo: Editora 34.

Hasenbalg, Carlos. 1979. *Discriminação e Desigualdades Raciais no Brasil.* Rio de Janeiro: Edições Graal.

Henriques, Ricardo. 2001. "Desigualdades Raciais no Brasil: Evolução das Condições de Vida na Década de 90". *IPEA (Texto para discussão 807)*: 1-49. Brasília/Rio de Janeiro: IPEA. http://www. ipea.gov.br/portal/images/stories/PDFs/TDs/td_0807.pdf

Htun, Mala. 2004. "From Racial Democracy to Affirmative Action: Changing State Policy on Race in Brazil." *Latin American Research Review* 39 (1): 60–89.

INCT – Instituto de Inclusão no Ensino Superior e na Pesquisa. 2013. *Mapa das Ações Afirmativas no Brasil.* http://www.inctinclusao.com.br/download/mapa_23maio2012cne.pdf

IPEA – Instituto de Pesquisas Econômicas e Aplicadas. 2012. *Políticas Sociais: Acompanhamento e Análise (20).* http://www.ipea.gov.br/portal/images/stories/PDFs/politicas_sociais/bps_20_completo.pdf

Lewandowski, Ricardo. 2013. *Arguição de Descumprimento de Preceito Fundamental 186 – O voto.* http://www.stf.jus.br/arquivo/cms/noticiaNoticiaStf/anexo/ADPF186RL.pdf

Moura, Clovis. 1977. *O Negro de Bom Escravo a Mau Cidadão?.* Rio de Janeiro: Editora Conquista.

Moura, Carlos, and Jônatas Barreto. 2002. *A Fundação Cultural Palmares na III Conferência Mundial de Combate ao Racismo, Discriminação Racial, Xenofobia e Intolerância Correlata.* Brasília: Fundação Cultural Palmares.

Munanga, Kabengele. 2003. "Políticas de Ação Afirmativa em Benefício da População Negra no Brasil: Um Ponto de Vista em Defesa das Cotas." In *Educação e Ações Afirmativas: Entre a Justiça Simbólica e a Injustiça Econômica*, edited by Petronilha Silva and Valter Silvério, 99–114. Brasília: Inep.

Nascimento, Abdias do. 1978. *O Genocídio do Negro Brasileiro: Processo de um Racismo Mascarado.* Rio de Janeiro: Paz e Terra.

Nascimento, Abdias do. 2003. *Quilombo: Vida, Problemas e Aspirações do Negro, 1948–1950.* São Paulo: Editora 34.

Osório, Rafael Guerreiro. 2003. "Mobilidade Social sob a Perspectiva da Distribuição de Renda." Master Thesis, Universidade de Brasília.

Paixão, Marcelo. 2006. *Manifesto Anti-racista: Idéias em Prol de uma Utopia Chamada Brasil.* Rio de Janeiro: Editoria DP&A.

Santos, Sales Augusto dos. 2003. "Ação Afirmativa e Mérito Individual". In *Ações Afirmativas: Políticas Públicas Contra as Desigualdades Raciais*, edited by Renato Emerson Santos, and Fátima Lobato, 73–92. Rio de Janeiro: DFP&A.

Santos, Sales Augusto dos2015. *O Sistema de Cotas para Negros da UnB: Um Balanço da Primeira Geração.* Paco Editorial: Jundiaí.

Silva, Tatiana Dias. 2012. "O Estatuto da Igualdade Racial. Brasília": *IPEA (Texto para discussão 1712)*: 1–66. http://www.ipea.gov.br/portal/images/stories/PDFs/TDs/td_1712.pdf

Soares, Sergei. 2008. "A Trajetória da Desigualdade: A Evolução da Renda Relativa dos Negros no Brasil." In *As Políticas Públicas e a Desigualdade Racial no Brasil: 120 Anos Após a Abolição*, edited by Mário Theodoro, 97–118. Brasília: IPEA.

Theodoro, Mario, ed. 2008. *As Políticas Públicas e a Desigualdade Racial no Brasil: 120 Anos após a Abolição.* Brasília: IPEA.

The challenge of creating a more diverse economics: lessons from the UCR minority pipeline project

Gary A. Dymski

ABSTRACT

This paper reflects on the experience of the 1999–2002 minority pipeline program (MPP) at the University of California, Riverside. With support from the American Economic Association, the MPP identified students of color interested in economics, let them explore economic issues affecting minority communities, and encouraged them to consider postgraduate work in economics. The MPP's successes and failures can be traced to the shifting balance in California's racialized political economy, especially a state ballot initiative forbidding the use of applicant race or ethnicity in University of California admission decisions, and to the transformation of economics itself, especially at the level of doctoral training. The MPP experience may be of relevance for other efforts to increase racial/ethnic diversity in social science disciplines.

1. Introduction

This paper reflects on the minority pipeline project (MPP) at the University of California, Riverside (UCR), sponsored by the American Economics Association between 1999 and 2002. The MPP identified students of color interested in engaging with economic issues affecting minority communities, let them research these issues, and encouraged them to pursue doctoral study in economics.

The ensuing discussion of the MPP offers some ideas on how to support minority students' critical engagement with issues of racial/minority inequality in the broader world. It also explores reasons why this program had limited success in its stated aim of increasing the flow of minority students into doctoral study in economics. This relative failure came despite the fact that UCR was able to maintain a racially diverse undergraduate student population, with relatively high levels of success in degree completion among minority

This paper originated as a presentation at the University of Leeds' 2013 Annual Black History Month Conference, *Building the Antiracist University: Next Steps*, which was sponsored by the University's Center for Ethnicity and Racism Studies (CERS) on 18 October 2013. The author thanks Dr Shirley Tate, CERS director, for her encouragement of – and patience during – the writing of this text. The reader is forewarned that the author has relied on memory to supplement notes taken during the period examined here. Any errors in this text are solely the author's responsibility.

students, in the face of mounting political pressures in California and at the federal level to disregard race in university admissions policies.[1]

We proceed as follows. Section 2 summarizes relevant aspects of the history of public policy in California, focusing on the implications of demographic shifts and economic growth for education policy. Section 3 describes the American Economic Association's evolving commitment to insuring racial/ethnic and doctrinal diversity. Section 4 then considers how UCR became a hub for racial/ethnic minority students in the 1990s and 2000s, even as legal challenges ended racial preferences in California's public education system. Section 5 describes the MPP itself. Section 6 concludes and draws tentative lessons from the UCR MPP experience.

Two caveats must be registered at the outset. First, the MPP experiment was undertaken largely without explicit institutional support. Second, the author's research as a heterodox economist focuses on how power, inequality, and uncertainty affect economic processes and outcomes: there is more interest in the historical trajectories and institutional settings of phenomena studied than in the behavioral interplay accompanying them. The emphasis would be reversed for a mainstream economist. The topic covered and they way they are approached – which follow lines of inquiry associated with figures such as Keynes, Marx, and Kalecki – are very different from those in most departments that grant doctoral degrees in economics in the US and UK.[2] Both these factors certainly shaped both the successes and failures of the MPP.

2. Growth and race in California: from Brown's master plan to proposition 13

Seventy years ago, Carey McWilliams described Southern California as an 'island on the land' (McWilliams 1946), due both to its location between mountains and ocean; his phrase also acknowledged the autonomy of California's political and economic trajectory relative to the rest of the nation. Influxes of new settlers in the 1800s, lured first by a gold rush and then by the completion of the transcontinental railroad, pushed aside the state's earlier inhabitants, both Native Americans and Mexicans alike, and built a growth economy geared to continual inflows of speculative capital and empire builders. With the onset of World War II, military bases and military-related manufacturing exploded in scale and size. By the war's end, California accounted for more than half of all US manufacturing capacity (McWilliams 1949); and the US economy as a whole, in turn, accounted for over half of all global manufacturing production. Postwar US programs encouraging home-ownership and creating modern highway systems rechanneled the state's furious growth rate toward suburban expansion. The future seemed without limit for this chronic growth economy. And if the cost of living hovered above that in the rest of the country, its schools were newly built, its teachers well-paid and newly appointed, and its street and freeway systems glittering and new.

It was widely accepted by the 1950s that this growing residential grid should be accompanied by an expanding public-education system; as Rarick (2005, 137) points out, 'Legislators believed, often correctly, that a college campus brought an area not only prestige but also jobs, innovation, and a better-trained workforce.' The rising political trajectory of Governor Pat Brown, who held office from 1958 to 1966, fed the state's hunger for more growth, as well as the expanded supply of water, food, and schools and universities this growth implied. Clark Kerr, taking the reins as University of California President, brokered the creation of a state higher education 'Master Plan,' which sought both to maintain academic standards and to continually expand access (Douglass 1999; Rarick 2005, 138–139, 147–153). The Plan, which quickly became a foundational reference-point in state political debates, dictated that the University of California have places for the top 12.5% of California high-school graduates in any given year, while the UC and California State systems together should offer places to the top third. This policy pre-ordained that the Riverside and Davis agricultural experimental stations would be converted into UC campuses, and that new public campuses would be built.

The surge of bond-fueled spending required to undertake the California Water Project and to expand highways and roads, public services, and educational campuses received the necessary legislative support in the early 1960s, the high water-mark of Brown's gubernatorial tenure. By 1966, social conflict and political strife opened the way for Ronald Reagan's successful campaign to replace Brown as California governor. Reagan emphasized law and order, with a double focus: first, to stand tall against Berkeley anti-war protesters; second, to take a hard line on race. The state had, since 1945, confronted the demand by minority advocates for fair employment and fair housing laws. These demands grew more vocal with the passage of the US Civil Rights Act in 1963. In 1964, California passed Proposition 14, opposing discrimination in housing (HoSang 2010, Chapter 3). Ten months later, the Watts race riots exploded the national myth of California as a land of dreamy surfers (Horne 1995). The Delano grape-workers strike, initiated in September 1965 by the United Farm Workers, followed by the March 1968 East Los Angeles walkouts by Chicana/Chicano students in the Los Angeles Unified School District, signaled that Brown Power, like Black Power, was coming of age.

While the bloom was not off the rose of the California myth, racial conflict had become a hardy perennial of state politics. The Vietnam War period pushed the state's military/ ethnic-growth spiral further along. It spurred the rapid growth of military-linked – and heavily Latino – cities such as Oceanside in Southern California and Vallejo and Martinez in Northern California; and it opened the way to the Immigration Act of 1965, which permitted a flood of legal immigration from Southeast and East Asia (initially Korea, the Philippines, Vietnam, and the Asian Chinese diaspora, and eventually China itself), as well as from Mexico and Central America. Further, the economic stress of the combined Wars on Poverty and on Vietnam, combined with growing global competition and 1970s spikes in global oil prices, led to several recessions – 1970, 1974, 1979, 1981 – which marked the end of what has been called the 'Golden Age' of capitalism (in the US and elsewhere in the global North (Glyn et al. 1990).

Combined with the declining pace of migration to California from other states within the United States, these changes combined to shift the terrain of white privilege: it was no longer the case, as in the early 1940s, that minority residents could only reside on the small percentages of California residential land not controlled by racial covenants; now white privilege was protected in property-rich enclaves. However, this barrier was itself eliminated by the California Supreme Court's 1971 and 1976 *Serrano vs. Priest* decisions. The Court ruled that K-12 education could no longer be funded by property taxes levied by individual school districts, as had previously been the case: the severe disparities in districts' property wealth led to violations of the US Constitution's equal-protection clause.

This decision, together with rising real-estate prices, soon fed a tax-revolt movement. In 1972, the amount of per-pupil property taxes was frozen in response to voter pressure (Kirst 2007). Six years later, a voter-passed Constitutional reform, Proposition 13, rolled back property taxes and 'froze' them at low levels.[3] Since property taxes were the foundation of school finances, this undercut K-12 education. A defining feature of Prop 13 was that when a home was sold, its property-tax burden would be reset in accordance with its higher property value.

The state government stepped in and made up the difference; but the primary financial consequence of Prop 13 was that education finance decisions were made, thereafter, at the state level, and thus were based on the state's general-fund revenues. The same had always been true regarding state support for California's public universities. The general fund itself depended primarily on income taxes. Due to California's progressive tax rates and to the variability of capital-gains income, these revenues were subject to severe boom-bust fluctuations.

Since the state's population was rapidly becoming more heavily minority – due to the impact of immigration from Asia and Latin America, in particular, and of differential birth rates for different demographic groups – a perpetual conflict was embedded in state politics: aging, largely white homeowners without school-age children, interested in low tax rates and law-and-order, vs. a younger, largely minority population with school-age children. Prop 13 had another key provision: it required a two-thirds vote of the California legislature to pass a budget and to pass a state tax increase. So state tax rates and local property-tax rates were both effectively locked in, as Prop 13 became a 'third rail' issue in California politics. Thus, in good fiscal years, spending might increase; in bad years, it tightened.

The result of these racially inflected, homeowner-newcomer conflicts in state finance was a systematic erosion in the fiscal resources available to public schools. While California had the 11th highest level of per-student K-12 expenditures among the 50 states in 1959–1960, its rank fell to 16th in 1969–1970, 22nd in 1979–1980, 32nd in 1989–1990, and 37th in 1999–2000.[4] After the subprime crisis, which hit California disproportionately hard, California's 2010–2011 per-pupil K-12 expenditures fell to 50th; only Utah spent less.[5]

This steady fiscal pressure on schools and educational resources had consequences. Parents who could afford to live in more expensive (and typically more white) communities with good public school systems moved there; private schools also proliferated. While the public University of California system retained its prestige, the affirmative action policies that had insured ethnically mixed entering classes were increasingly challenged. Most famously, the US Supreme Court ruled, in its 1978 Board of Regents vs. Bakke decision, that while race could be considered as one consideration in admission decisions, racial quotas were impermissible.

3. Diversity and American academic economics

The economic and social macrodynamics that led California from postwar building phase to tax revolt in less than two decades has had similarly dramatic effects on the academic economics profession. The diversity of theoretical perspectives in academic economics was profoundly affected, as was the representation of racial/ethnic minorities within the profession.

From the mid-1960s to the mid-1970s, significant space was provided in mainstream journals for dialogs between economists holding 'mainstream' views and those expositing heterodox approaches rooted in the ideas of Marx and Keynes. An extended controversy over the treatment of production in aggregate models, featuring mainstream proponents disproportionately based in Cambridge, Massachusetts, and heterodox economists linked to Cambridge, England, raged for a decade (Harcourt 1969). This 'Cambridge controversy over capital' was not an isolated case: for example, Crotty and Rapping (1975) published a no-holds-barred left critique of the federal budget in the flagship *American Economic Review* in the mid-1970s; and Harrison (1972, 1974a, 1974b) published a series of papers on the ghetto economy (that is, the economic dynamics of the segregated inner-city economy) in top-ranked mainstream journals.

However, this window of academic exchange for heterodox economists in mainstream journals soon closed. The steady growth and low inflation of the 1950s and 1960s was replaced by stagnation and crisis. This legitimized orthodox articles demonstrating that 'Keynesian consensus' macroeconomic models led to misguided over-reliance on state policies; liberating market forces and corralling the state was needed to renew growth. Any interest that more tolerant mainstream economists had had in dialogs with Keynesian and neo-Marxian models challenging the premises of equilibrium-based economics largely evaporated. After the late 1970s, analyses of 'real world' developments from a Keynesian/Marxian perspective that did not prioritize mathematical modeling or econometric results could be published, in the main, only in heterodox journals.

Meanwhile, the inattention of the economics profession to issues affecting minorities drew a response. In 1969, Robert S. Browne founded the Black Economic Research Center in Harlem; its advisors and staff undertook policy-oriented research focused on reparations and Black economic development, among other issues. In that same year, the Caucus of Black Economists was founded, and the *Review of Black Political Economy* (RBPE) was launched, edited by Dr Browne. His founding statement envisioned the RBPE as 'an hospitable arena in which black people could explore ideas as to how they might bring about effective and substantial improvement in their collective economic position' (Betsey 2008, 36).

In the following year, the Caucus of Black Economists initiated discussions with the American Economic Association executive board about the paucity of minority economists (Collins 2000).[6] This led to the founding in 1974 of the Committee on the Status of Minorities in the Economics Profession (CSMEP). Its goal was to increase the representation of minorities among academic economists, with the twin purposes of improving economists' policy advice and providing more role models for young minority students. As Collins (2000) documents, the percentage of (US national) minority students receiving bachelors' degrees is below that for white students, and the percentage of minorities receiving doctorates is half of this low level (and has remained there from the 1970s to the 1990s); and the representation of minorities in academic positions is a third of that low level (about 1–2% of the total, for US-national faculty members).

That same year, CSMEP launched the first annual Summer Program, with the aim of preparing aspiring minorities for doctoral study in economics. By 1998, the CSMEP's Summer Program had hosted 634 students. Collins (2000) found data for 348 of these students. Of this total, 129 had started a PhD program; 46 had completed their doctorates; 56 were currently enrolled, and 27 had left without earning a doctorate. To put these figures into context, the 3990 faculty members in US academic economics departments as of 1998 included only 150 Black and Hispanic members (Collins 2000, Table 7).

4. Riverside: after the end of affirmative action, the rise of a racially diverse campus

Since its conversion from an agricultural demonstration station in 1954, the Riverside campus of the University of California has held a distinctive place within the system. Riverside was originally envisioned as an experiment – a small liberal arts college inside the UC system, a 'Swarthmore of the West' (Agha 2004). This experiment soon gave way to pressures stemming from population growth, which triggered University expansion under the Master Plan. To relieve pressure on the Berkeley and Los Angeles campuses, Davis and Riverside were converted into general campuses, and several new UC campuses – Santa Cruz, Santa Barbara, Irvine, and San Diego – were built in attractive seaside locations. UCR struggled, as its smoggy inland site made it the least attractive general campus option for UC-eligible students.

4.1. An end to affirmative action

Proposition 13 became law in California in the June primary election of 1978. While Democratic Governor Jerry Brown (son of Governor Pat Brown) was re-elected in November 1978, his colorful personal life and his political agenda – including advocacy of environmental sustainability and of the creation of a state space agency – were out of step with a political shift toward 'traditional values' and nationalism. The national re-alignment, marked by the 1980 Presidential victory of former California governor Ronald Reagan, was followed in California by 1982 and 1986 election mandates for pro-business Republican Governor George Deukmajian. In 1990, the Republican mayor of San Diego, Pete Wilson, was elected California governor on the strength of a law-and-order agenda that included the rollback of programs benefiting minorities and undocumented immigrants. Wilson appointed Ward Conerly to the Board of Regents of the University of California in 1993. Conerly, a politically connected small-business owner whose firm had taken advantage of governmental affirmative-action programs over the years, was an outspoken opponent of affirmative action programs (Musgrove 1999).

Wilson's successful 1994 re-election campaign featured his support for Proposition 187, which prohibited undocumented immigrants' access to health, education, and social services.[7] In 1995 Conerly won Regental support for ending affirmative action in the University of California. Governor Wilson, seeking to bolster his bid for the 1996 Republican Presidential nomination, supported Proposition 209. This California referendum sought to ban the use of race, sex, ethnicity or national origin – that is, affirmative action – in the provision of public services. Ward Conerly campaigned fiercely for the measure. It passed 54–46 in November 1996, at least in part because many Californians voting for the California Civil Rights Initiative mistakenly thought their 'yes' votes registered support for, not opposition to, affirmative action (Musgrove 1999).

These broader political trends deeply affected public higher education generally, and the UC system and UCR in particular. The *Bakke* decision effectively opened the floodgates for further legal attacks on affirmative action programs aimed at reducing racial disparities in university admissions (Solórzano and Yosso 2002). Ethnic studies programs across the country, targeted by well-funded campaigns (Nicol 2013), were eliminated or consolidated. At UCR, what had been several distinct departments (Black Studies, Chicano Studies, and

Native American Studies) were merged into one 'ethnic studies' department in 1981. The Regents' 1995 anti-affirmative action resolution and the Proposition 209 campaign, publicly led by a UC Regent and supported by private donors, was broadly resisted by UC faculty and students (Taylor 1999). The implementation of Proposition 209, once it was passed, had a devastating effect on the number of minorities admitted to UC law and medical schools (Karabel 1999; Moran 2000).

4.2. The 'tidal wave' comes to UC riverside

A collision between the political mandate to end affirmative action, the tax revolt, and the demographic bargain built into the Master Plan was inevitable. Popular and political support was mobilized to insure that the 'tidal waves' of prospective higher education students entering the system from the early 1990s onward – the children of the US 'baby-boomer' generation and of immigrant households – would find places in the state's public universities.[8] Sustaining this bargain even as Prop 209 bit home meant raising enrollment fees; for the UC system specifically, it also meant finding campuses with growth capacity. Riverside volunteered itself for this expansion.

The Riverside campus had languished in its post-'Swarthmore of the west' period; rumors circulated that it was targeted for closure. Embracing 'Tidal Wave' growth insured survival. The campus's location 55 miles to the east of downtown Los Angeles put it near the center of population growth in Southern California – both the African-American and Chicano-Latino hubs of South Central and East Los Angeles, and the rapidly expanding Asian-American 'ethnoburbs' (Li 2008) of the San Gabriel Valley. Consequently, Riverside met its Master Plan commitments by drawing heavily on this largely minority and immigrant population base. Given the complex family demands on minority/immigrant students, their families' lower-than-average incomes, and rising UC tuition, most of UCR's growth came from minority commuter students.

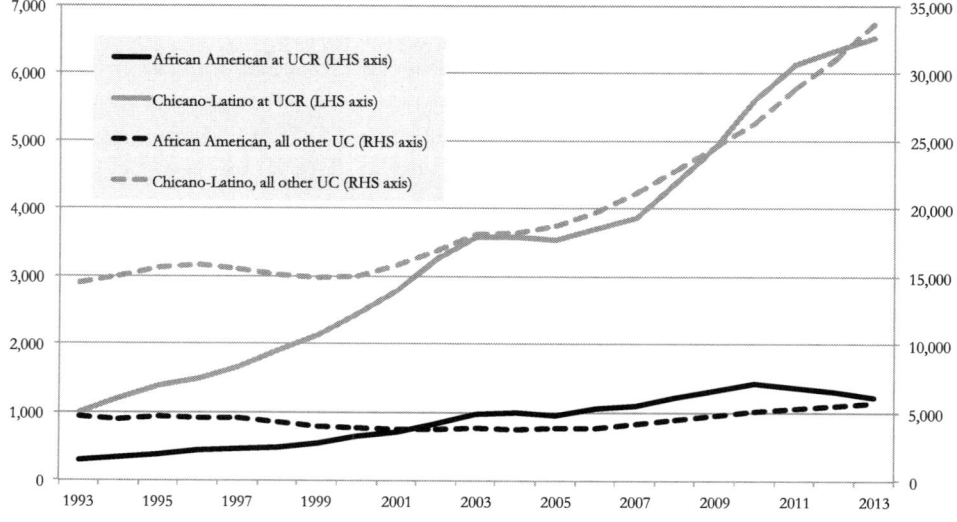

Figure 1. African-American and Chicano-Latino undergraduates enrolled, UC Riverside vs. other UC campuses, 1993–2013.

Figures 1–4 depict various aspects of UC undergraduate enrollment growth between Fall 1993 and Fall 2013. Figure 1 contrasts the rising numbers of African-American and Chicano-Latino students at UCR with those at the other UC campuses.[9] Whereas the number of African-American students in other UC campus declines by 18% between 1993 and 2004, it increased by 240% at UCR; the changes in Chicano-Latino enrollments are, respectively, increases of 25 and 259%. Figure 2 shows Asian-American enrollments rising rapidly, both at UCR and at other UC campuses (the respective percentage increases for the 1993–2004 period are 53 and 151%). This figure also illustrates that white student enrollments remained relatively constant (0.3 and 3.4% between 1993 and 2004, respectively).

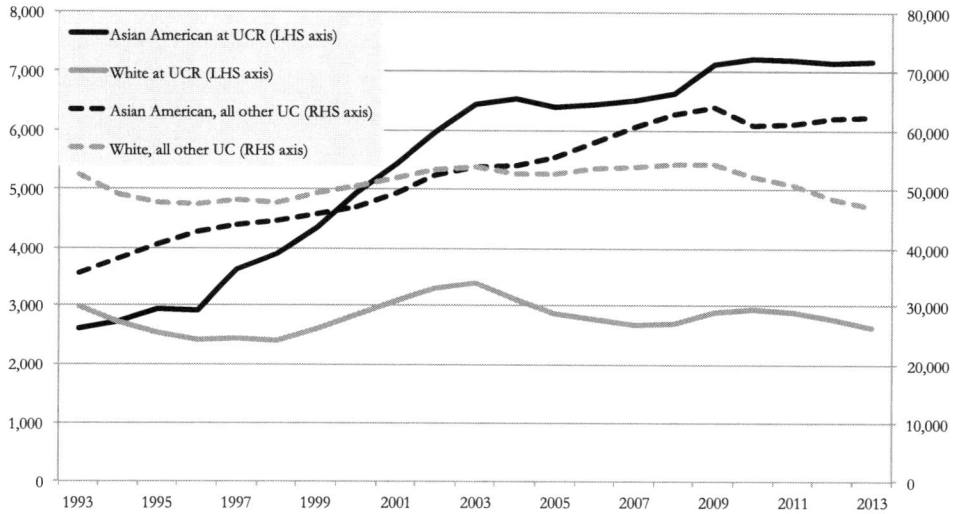

Figure 2. Asian-American and White undergraduates enrolled, UC Riverside vs. other UC campuses, 1993–2013.

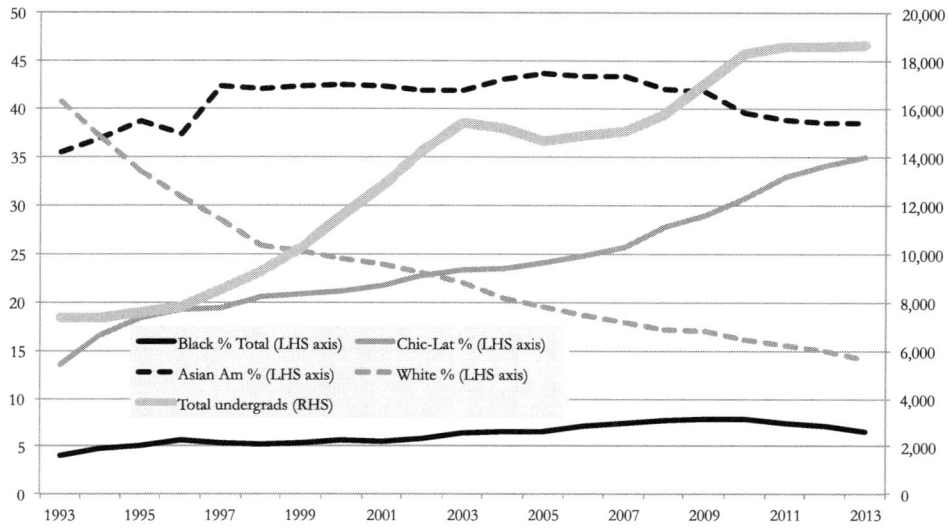

Figure 3. Undergraduate enrollment percentages by ethnicity plus enrollment total, UC Riverside, 1993–2013.

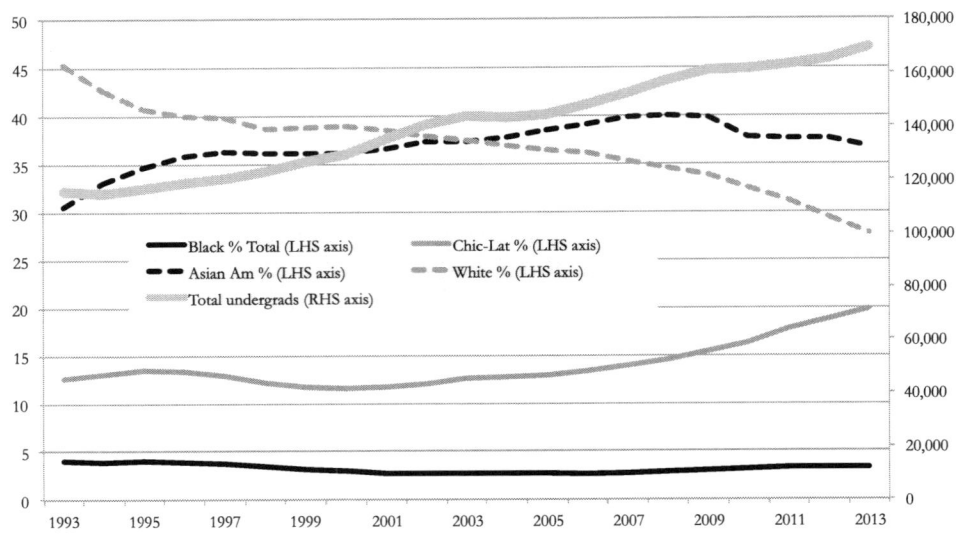

Figure 4. Undergraduate enrollment percentages by ethnicity plus enrollment total, all UC campuses except Riverside, 1993–2013.

Note that while the UCR/Other-UC data are depicted at a scale of 1:5 in Figure 1, they are shown at a scale of 1:10 in Figure 2. This demonstrates the 'minority-focused' nature of Riverside's campus. Figures 3 and 4 contrast the rapid rising percentages of minority students (and declining share of white students) at UCR with the more moderate trends at other UC campuses; these figures also depict overall enrollments.

Thus, despite the legal restrictions imposed on the UC system's ability to recruit for ethnic/racial diversity, its undergraduate population became more ethnically diverse. This was especially so for UCR, which was among the 'non-selective' campuses in the system.[10] In the 2001 *U.S. News and World Report* computations, UCR ranked as the third most ethnically and racially diverse public university in the U.S. The expected implication of taking on a heavily minority student body with a family income $10,000 less than for other UC campuses is that Riverside would be the least among equals. One would further anticipate that campus morale might plunge and student drop-out and time-to-degree rates might climb.

UCR did not, however, slink down this path. Keyed by the 1992 arrival from UCLA of UCR chancellor Raymond Orbach, it embraced its future as a multi-cultural, multi-ethnic institution. Chancellor Orbach made two immediate moves: he raised performance standards for UCR faculty and personally led an aggressive, all-out recruiting drive in the predominantly minority, lower income, heavily immigrant high schools of the eastern portions of Los Angeles. If these schools would institute more demanding curricula that complied with UC admissions standards, their 'best and brightest' students could apply to UCR. Second, UCR maintained and continued to invest in its distinct student-support and activities centers; there are distinct academic/social hubs for African-American, Latino/Chicano, Asian-American/Pacific Islander, Native American, and international students, each run by a full-time director assisted by a small staff. Providing distinct 'homes' permits student leaders and student-services staff to utilize culturally nuanced encouragement to students, providing social and emotional validation and helping facilitate successful competition in the academic sphere. The result is that minority students in a diverse multi-cultural context are made to feel welcome on campus. Informal evidence collected by UCR admissions counselors indicates that this ethnic mix and the campus's reputation as being supportive of minority-student presence has helped make Riverside a destination campus for many minority applicants to the UC system.[11]

5. The minority pipeline program at UCR

Just as UCR enrollments entered a period of precipitous growth, I left my former academic position at the University of Southern California (USC) and joined the economics faculty at UC Riverside. In my five years at USC, I had compiled a list of publications, as the 'publish or perish' system demands. These were all, however, in heterodox economics journals or journals outside of economics; as such, they would not suffice for a successful tenure application to USC's ambitious mainstream economics department; UCR's economics department, on the other hand, had become a national beacon for heterodox economics.

In 1990 and 1991, some collaborators and I had undertaken an extensive report for LA Mayor Tom Bradley and the City Council on racial and income inequality and banking in the city. The report was released in early November 1991. Just under six months later, the city burned. From 29 April to 4 May 1992, the LA uprising – triggered by judicial inaction in response to the taped beating of Rodney King – devastated parts of the city.[12] Many public forums were held in subsequent weeks throughout the city; as one result of such events, scholars such as Eugene Grigsby, Melvin Oliver, and Paul Ong made major advances in linking economic dynamics, racial inequality, and urban social movements.

All these events created a sense of urgency about racial/ethnic inequality in Southern California, then, just as the UC student 'Tidal Wave' landed at UCR. In Fall 1994 and Fall 1995, I offered an undergraduate module called 'The Political Economy of Los Angeles.' I also sponsored four students in summer 1994 under the auspices of the UCR-funded Minority Summer Research Internship Project (MSRIP). Two of these modules' students went to the CSMEP Summer Program – and one of those two, a Chicano/Latino student, went on to receive a doctorate in economics from Duke University.[13] While offering this summer program, I made contact with Professor Cecilia Conrad of Pomona College. Cecilia, a CSMEP board member (and now editor of the RBPE), wanted to support some experiments in building the pipeline of minority economics students.

We agreed on trying one such program at UCR, due to its growing diversity (Figures 1–4), its reputation for supporting minority students' success, and the fact that UCR had sent students on to both the CSMEP Summer Program and even graduate study in economics. The premise of the experiment was that many students who might otherwise gravitate toward economics were discouraged by the formalism and abstract content of introductory modules.[14] Why not, then, contact students in creative ways at formative stages in their undergraduate careers?

The UCR MPP's first year of operation was 1999–2000. Presentations about the MPP were made in open meetings held at each of the four minority-student activities centers, in the economics department, and at the UCR honors program. A speaker series was initiated, as were study groups and some innovative methods of academic staff/student contact:

1. In four different programs, invited speakers spoke about their own life experiences in economics in interactive sessions, presented formal seminars, and then met informally with students and faculty. These guests included two Latino economists

(Ron Oaxaca from University of Arizona and Arturo Gonzalez from Arizona State University), an African-American economist (Curtis Haynes from Buffalo State University), and an Asian-American economist (Paul Ong from UCLA). These visits, co-sponsored by the minority student program centers, received enthusiastic support from these centers' staffs and students.

2. Economics study groups were initiated for a course in intermediate macroeconomics taught in Spring 2000 (Conrad and Sharpe 1996; Karabel 1999; Moran 2000). While aimed at minority students, these groups were open to all participants. They enabled participating students to improve their worksheet and exam performance. Further, the UCR Learning Center trained special peer-group course counselors; these counselors' expenses were funded by the MPP.

3. An experimental approach to faculty-student contact was tried out for the Spring 2000 intermediate macroeconomics course mentioned above. After the 225-student Tuesday-Thursday lecture concluded, anywhere from 3 to 35 interested students convened at patio tables in the UCR Commons area for informal two-hour sessions. Topics ranged from module concepts to mentoring to discussion of professional experience. While all students were welcomed at these sessions, the MPP was advertised. Minority students were disproportionately represented at these sessions. Among the attendees were two members of UCR's varsity basketball team.[15]

In the second year of implementation, academic year 2000–2001, these activities were sustained.[16] In particular, item 2 was augmented. Learning Center staff institutionalized peer-counseling study groups for microeconomics and macroeconomics. These counselor and study-group sessions for economics modules were subsequently built into the Learning Center's operating budget.[17] In addition to the activities named above, three more were added: two innovative undergraduate courses open to non-majors; a distinguished guest-scholar visit to UCR, sponsored by the MPP; and a summer research initiative.

The innovative course offerings in 2001 and beyond included the re-launching of Economics 146, Urban Economics. Like the Honors module, Economics 146 took a 'hands on' approach. In 2001, for example, most students participated in a community-based project in conjunction with the UCR-affiliated Community Digital Initiative (CDI) at the Cesar Chavez Center. The class undertook a multi-dimensional study of the CDI's IT training program for at-risk youth, which had the aim of placing these youth in local firms. Economics 146 students met with participating youth, interacted with staff, assessed local secondary and post-secondary IT-training programs, and conducted surveys of both IT-related and non-IT small businesses. The study reached a pessimistic conclusion – there was little evidence of a latent or real demand for these IT-trained youth's labor. The project did succeed in interesting some Economics 146 students in the challenges of urban economic development.

In that same academic year, the UCR Honors Program agreed to sponsor an honors module on racial inequality. This Winter 2001 module, aimed at non-majors with no background in economics, was opened to UCR students beyond those in the Honors program. Approximately 24 students, disproportionately minority, enrolled in this course in each of the three successive years in which it was offered. Module students implemented 'hands on' tests of the presence of racial inequality in the market economy.

During May 2001, James Stewart, a professor and former vice provost for educational equity at Pennsylvania State University, spent a week at UCR as a scholar in residence.

Professor Stewart, formerly editor of the RBPE and President of the National Economics Association was, in that period, president of the Association for Black Studies. He had numerous meetings with university executives, economics and academic staff, and students. He gave lectures and presentations for the student Economics Honors society, California State University, San Bernardino (CSUSB), Economics 146, and the African Student Alliance. His visit galvanized many students' interest in economic issues affecting minority communities.

The final event in the 2000–2001 academic year was a summer research project. Minority students identified through MPP events were invited to participate in a special summer research initiative. With support from the MSRIP program, this initiative provided a first-hand research experience permitting undergraduates to connect economics concepts with the circumstances of communities like those in which they had grown up. Seven students were selected for participation via a screening process: three through the MSRIP, and four through the resources of the MPP. These students included four African-Americans, one student whose parents are African-American and Native American, one Asian-American, and one international Asian student. In addition, six students participated on a volunteer basis: one Latino student, one African-American student, and four Asian-Americans. Helping in the coordination of this project was Professor Carolyn Aldana of CSUSB, a Latina, who had earned her doctorate in economics from UCR in 1995.

This lively and motivated cohort of students took a large share of responsibility for designing and implementing their research project. They assessed the access of minority-owned small businesses in the City of Riverside and in a portion of the eastern San Gabrial Valley to customers, to suppliers, and to credit. The project unfolded over 8 weeks during summer 2001. The students succeeded in obtaining 168 completed surveys, providing the basis for a co-authored paper, which was completed in the Fall 2002 quarter. While the MPP ran for a third year, we turn now to an assessment.

6. Conclusion: an assessment of the UCR MPP

The MPP posed a question: can classroom and non-classroom methods be used to motivate minority students to *consider* doctoral study in economics? The answer is unambiguously yes. It also increased many students' proficiency in economics, and underwrote several collectively planned and implemented research projects. Several MPP participants went into the CSMEP Summer Program. One MPP participant became the first male African-American valedictorian at a UCR graduation (see footnote 15).

But did the MPP succeed in *producing* more economists? No. One participant from the three years received a degree in agricultural economics; another received a doctorate in public policy (and is now a professor in the public New York university system). Why not? One reason is that many participants in the MPP had heavy family obligations, including the provision of financial support, which blocked any thought of doctoral study. Other participants came from families without university experience and with ideas about their sons' or daughters' future trajectories which did not include graduate study.

Two MPP members were accepted into economics doctoral programs, and spent a year or more therein. However, they did not stay on; others did not try this path at all. Undoubtedly, the heavily formalistic and mathematical training of US economics doctoral training constituted a double barrier: first, all the MPP participants had graduated from California public schools, whose systematic defunding has been documented above; second, the community-focused work that the MPP students undertook is little valued in economics doctoral training. In effect, the very work that had led students to consider economics would push them away once they got there.

Another factor in the MPP's failure to generate more economics doctorates has to be laid to the guidance provided. First, the author did not succeed in recruiting colleagues to join in this program, both because no financial incentives were available, and these colleagues generally had no experience in community-based work. Second, the author's own research interest imposed limits. The topics highlighted in MPP activities and modules stood at some distance from that of mainstream economics – which is what one encounters in required doctoral coursework. In effect, MPP students were being pulled toward perspectives more common in sociology or geography than in economics graduate programs. Table 6 in Collins (2000) sets out the academic fields in which minority and non-minority economists work: political economy, one of the author's fields, is not even listed as a topic; and urban/regional economics encompasses a total of only nine minority economists nationwide.

Academic studies published after the termination of the MPP suggest that encouraging student interested in the economic aspects of the communities like those they had grown up is not a common pathway into the profession. Stock and Siegfried (2015) document that since 1982, foreign and not domestic students have received the majority of economics PhDs from US universities (currently, the figure stands at 55%). And Evans, Grimes, and Becker (2012), in an examination of what led eminent economists to enter the field, find that few were engaged with economics as undergraduates.

But this underlines, rather than undercutting, the importance of linking research in economics to the economic and social dynamics that affect minority communities. The point of training community-minded students as economists is to ensure that work which renders the structure of racial inequality visible is done *within* the domain of economics, rather than being possible only outside of it. The importance of this cannot be exaggerated, at a time when the public policy trend toward invisibilizing white privilege in color-blindness continues (as the passage of the 'No Child Left Behind' Act of 2001 illustrates (Leonardo 2007)).

That having informed participants – those who have experienced urban inequality in the US, for example – at the analytical table matters is evident by registering the change in focus in economists' writing about credit and race in a short space of time. In the 1990s, Munnell et al. (1996), economists in the Federal Reserve system, produced evidence demonstrating the existence of racial redlining in Boston, in a series of path-breaking papers. This work, in effect, critiqued the structure of racial inequality. But when four Federal Reserve economists wrote a definitive study of the causes of the subprime crisis (Gerardi et al. 2008), they focused primarily on the existence of neighborhood spillover effects in housing prices – the structure of racial inequality was off their radar. Even while subprime loans were being introduced into the marketplace, two different papers used very similar analytical frameworks based on asymmetric information in a principal-agent setting to come to very different conclusions: Calomiris, Kahn, and Longhofer (1994) argue that credit-market redlining is a legitimate practice because those who are redlined, if they were *not* to be redlined, *would* default at higher rates; and Dymski (1995), who condemns this practice as unjust on its face. This inattention to the socio-economic dynamic of race encourages work about economic issues imbued with racial dynamics which pushes race even further into the background; as an example, consider the paper by Ambrose, Capone, and Deng (2001) which considers mortgage foreclosures as 'put options' for homeowners. It comes as no surprise to see Calomiris and Haber (2015) subsequently blame government interference in market processes for the subprime crisis.

There is something broken about a profession whose most respected experts almost invariably blame government interference for economic policy failures and crises, with the consequence that government policy-makers have even less scope for halting the chaos that can result when market forces are freed of any constraint. It cannot be too much to hope that those within the discipline can remake it as a tool for taking on, rather than invisibilizing, the social divisions that stand in the way of a more widespread prosperity.

Notes

1. The expression 'K-12' denotes 'Kindergarten to 12th grade' – that is, the entirety of 'grade school,' 'middle school,' and 'high school' in the US system. Students completing grade 12 can either terminate their schooling, enter vocational programs, or enroll in a university.
2. Dymski (2014) compares heterodox and mainstream approaches to economics.
3. Since 1911, California has had an initiative procedure wherein proposed changes to the state's constitutions can be considered as ballot propositions in general elections. Passage requires only a simple majority of all those voting. This Populist Era mechanism, originally passed to empower grassroots sentiment, now regularly attracts millions of dollars from the 'third house' (Michael and Walters 2002).
4. The source of these statistics is Table 194 in the *Digest of Educational Statistics, 2010*, US Department of Education. Washington, DC, 2010. See https://nces.ed.gov/programs/digest/d10/tables/dt10_194.asp
5. See John Fensterwald, 'Latest – but outdated – Ed Week survey ranks California 50th in per pupil spending,' *EdSource*, January 13, 2014. Accessed at http://edsource.org/2014/latest-but-outdated-ed-week-survey-ranks-california-50th-in-per-pupil-spending/56196.
6. As Collins notes, 'minority' was defined to include Blacks, Hispanics, and Native Americans.
7. Proposition 187 was ruled unconstitutional by a federal appellate court in 1987; Democratic Governor Gray Davis chose not to appeal the ruling, and the measure was never implemented.
8. See California Citizens Commission on Higher Education (1999), Bracco and Callan (2002), and http://www.ucop.edu/acadinit/mastplan/.
9. Native American students are excluded from Figures 1–4 due to their small numbers (as are international students and students not specifying their ethnicity). Note that the enrollment of Native American students at UCR increased 44% between 1993 and 2004, while at other UC campuses it fell by 27%.
10. Students meeting the 'Master Plan' threshold were assured of a place in the UC system, but not at the campus of their choice. So while some campuses could select from among a surplus of applicants, other campuses would take students who were UC-qualified but whom other campuses had not accepted.
11. Ball, Reay, and David (2002) provide empirical evidence that campus ethnic mix affects minority students' choices among higher education institutions. UCR's strength in attracting and retaining students of color has led to its being ranked first in 2014 in the *Washington Monthly* poll of US universities meeting the 'Obama criteria' of access, diversity, affordability, and success probability.
12. Davis (1990) provides an unflinching view of the social and economic conditions that led to the 1992 uprising.
13. The other student became the first African-American to be named valedictorian for a graduating class at UCR. Despite receiving high marks in doctoral courses in microeconomics while completing her undergraduate degree, she entered a teachers-education program, and now teaches in an elementary school in Atlanta.

14. Note that in the US system, students enter universities without declared majors, and instead sample courses across a range of disciplines before choosing a major in their second year of study.
15. One of these students, Zack Elder, participated in the subsequent MPP summer program and went on to earn his doctorate in economics. He now lives and teaches in Asia.
16. Guest lectures were given by Michael Stoll of UCLA and by Jessica Gordon Nembhard of the City University of New York.
17. This institutionalization was spearheaded by Teresa Cofield of the Learning Center. Ms Cofield, entered the doctoral economics program at Pennsylvania State University after graduating from UCR in 1993 with a double major in economics and mathematics. After one year, she returned to Southern California, subsequently earning an EDD degree.

Disclosure statement

No potential conflict of interest was reported by the author.

References

Agha, Marisa. 2004. "Celebration: The University is Marking its 50th Anniversary with a Variety of Events." *Riverside Press-Enterprise*, January 19.

Ambrose, Brent W., Charles A. Capone, Jr., and Yongheng Deng. 2001. "Optimal Put Exercise: An Empirical Examination of Conditions for Mortgage Foreclosure." *The Journal of Real Estate Finance and Economics* 23 (2): 213–234.

Ball, Stephen J., Diane Reay, and Miriam David. 2002. "'Ethnic Choosing': Minority Ethnic Students, Social Class and Higher Education Choice." *Race Ethnicity and Education* 5 (4): 333–357.

Betsey, Charles L. 2008. "A Brief Biography of Robert S. Browne." *The Review of Black Political Economy* 35: 57–60.

Bracco, Kathy Reeves, and Patrick M. Callan. 2002. "Competition and Collaboration in California Higher Education." *Report No. 02-1*. San Jose, CA: National Center for Public Policy and Higher Education.

California Citizens Commission on Higher Education. 1999. *Toward a State of Learning: California Higher Education for the Twenty-first Century*. Los Angeles, CA: Center for Governmental Studies.

Calomiris, Charles W., and Stephen H. Haber. 2015. *Fragile by Design: The Political Origins of Banking Crises and Scarce Credit*. Princeton, NJ: Princeton University Press.

Calomiris, Charles W., Charles M. Kahn, and Stanley D. Longhofer. 1994. "Housing-finance Intervention and Private Incentives: Helping Minorities and the Poor." *Journal of Money, Credit and Banking* 26 (3), Part 2: 634–674.

Collins, Susan M. 2000. "Minority Groups in the Economics Profession." *Journal of Economic Perspectives* 14 (2): 133–148.

Conrad, Cecilia A., and Rhonda V. Sharpe. 1996. "The Impact of the California Civil Rights Initiative (CCRI) on University and Professional School Admissions and the Implications for California Economy." *The Review of Black Political Economy* 25 (1): 13–59.

Crotty, James R., and Leonard A. Rapping. 1975. "The 1975 Report of the President's Council of Economic Advisors: A Radical Critique." *American Economic Review* 65 (5): 791–811.

Davis, Mike. 1990. *City of Quartz*. London: Verso.

Douglass, John Aubrey. 1999. "The Evolution of a Social Contract: The University of California before and in the Aftermath of Affirmative Action." *European Journal of Education* 34 (4): 393–412.

Dymski, Gary A. 1995. "The Theory of Bank Redlining and Discrimination: An Exploration." *The Review of Black Political Economy* 23 (3): 37–74.

Dymski, Gary A. 2014. "The Neoclassical Sink and the Heterodox Spiral: Political Divides and Lines of Communication in Economics." *Review of Keynesian Economics* 2 (1): 1–19.

Evans, Brent A., Paul W. Grimes, and William E. Becker. 2012. "What Led Eminent Economists to Become Economists?" *The Journal of Economic Education* 43 (1): 83–98.

Gerardi, Kristopher S., Andreas Lehnert, Shane M. Sherland, and Paul S. Willen. 2008. "Making Sense of the Subprime Crisis." *Brookings Papers on Economic Activity* Fall: 69–145.

Glyn, Andrew, Alan Hughes, Alain Lipietz, and Ajit Singh. 1990. "The Rise and Fall of the Golden Age." In *The Golden Age of Capitalism: Reinterpreting the Postwar Experience*, edited by Stephen Marglin and Juliet Schor, 39–125. Oxford: Clarendon Press.

Harcourt, Geoffrey C. 1969. "Some Cambridge Controversies in the Theory of Capital." *Journal of Economic Literature* 7 (2): 369–405.

Harrison, Bennett. 1972. "Education and Underemployment in the Urban Ghetto." *American Economic Review* 62 (5): 796–812.

Harrison, Bennett. 1974a. "Ghetto Economic Development: A Survey." *Journal of Economic Literature* 12 (1): 1–37.

Harrison, Bennett. 1974b. "Ghetto Employment and the Model Cities Program." *Journal of Political Economy* 82 (2), Part 1: 353–371.

Horne, Gerald. 1995. *The Fire this Time: The Watts Uprising and the 1960s*. Charlottesville, VA: University of Virginia Press.

HoSang, Daniel Martinez. 2010. *Racial Propositions: Ballot Initiatives and the Making of Postwar California*. Berkeley: University of California Press.

Karabel, Jerome. 1999. "The Rise and Fall of Affirmative Action at the University of California." *The Journal of Blacks in Higher Education* 25: 109–112.

Kirst, David. 2007. *The Evolution of California's State School Finance System and Implications from Other States*. Stanford, CA: Institute for Research on Education Policy and Practice, Stanford University.

Leonardo, Zeus. 2007. "The War on Schools: NCLB, Nation Creation and the Educational Construction of Whiteness." *Race Ethnicity and Education* 10 (3): 261–278.

Li, Wei. 2008. *Ethnoburb: The New Ethnic Community in Urban America*. Honolulu: University of Hawaii Press.

McWilliams, Carey. 1946. *Southern California: An Island on the Land*. Layton, UT: Gibbs Smith.

McWilliams, Cary. 1949. *California: The Great Exception*. Berkeley: University of California Press.

Michael, Jay, and Dan Walters. 2002. *The Third House: Lobbyists, Money, and Power in Sacramento*. Berkeley: Public Policy Press.

Moran, Rachel F. 2000. "Diversity and its Discontents: The End of Affirmative Action at Boalt Hall." *California Law Review* 88 (6): 2241–2352.

Munnell, Alicia H., Geoffrey M. B. Tootell, Lynn E. Browne, and James McEneaney. 1996. "Mortgage Lending in Boston: Interpreting the HMDA Data." *American Economic Review* 86: 25–53.

Musgrove, George Derek. 1999. "Good at the Game of Tricknology: Proposition 209 and the Struggle for the Historical Memory of the Civil Rights Movement." *Souls* 1 (3): 7–24.

Nicol, Donna J. 2013. "Movement Conservatism and the Attack on Ethnic Studies." *Race Ethnicity and Education* 16 (5): 653–672.

Rarick, Ethan. 2005. *California Rising: The Life and times of Pat Brown*. Berkeley: University of California Press.

Solórzano, Daniel G., and Tara J. Yosso. 2002. "A Critical Race Counterstory of Race, Racism, and Affirmative Action." *Equity & Excellence in Education* 35 (2): 155–168.

Stock, Wendy A., and John J. Siegfried. 2015. "The Undergraduate Origins of PhD Economists Revisited." *The Journal of Economic Education* 46 (2): 150–165.

Taylor, Ula. 1999. "Proposition 209 and the Affirmative Action Debate on the University of California Campuses." *Feminist Studies* 25 (1): 95–103.

Dealing with difficult conversations: anti-racism in youth & community work training

Diana Watt

ABSTRACT

This paper represents a critical reflection on youth and community work students' response to a module on race equality and diversity. An awareness of issues in relation to power and oppression are amongst the core elements of youth and community work training. Throughout their study, youth and community work students are engaged in conversations aimed at enabling them to critically examine their own attitudes and beliefs in areas of anti-discriminatory and anti-oppressive practices. These classroom conversations and expressions of resistance and resilience are informed by Paulo Freire's work on critical dialogue. As a specialist unit the module on equality and diversity was aimed at developing students' critical understanding of race, racism and ethnic difference. Based on written feedback, student-led presentations and conversations of 'protest', this paper critically explores the power of whiteness in silencing particular groups of students.

Introduction

Youth and community work in the UK is a contested area of professional practice and this is reflected in the range of degree titles and joint awards. Existing courses include: Community and Youth Work, Community Education, Community and Youth Studies, Community Learning and Development, Informal Education, Community Youth Work and Community Development, Youth Work and Youth Studies, Youth Work and Sports Science, Childhood and Youth Studies (http://view.qaa.ac.uk/Publications/Documents/Subject-benckmark-statement-youth, accessed 20 May 2016). Irrespective of the differences in course titles, the pedagogic emphasis is on conversation and dialogue. When compared to monologue, the dominant form of communication in schools, conversation and dialogue is that which is seen as giving voice to those who have been silenced (Batsleer 2008, 5). Conversation is thus described as the 'stock and trade' of youth and community work practice (Soni 2011). Most conversations are usually concerned with issues of power, inequality, difference, participation and transformation. As a strategy for dealing with racism at an institutional level and improving levels of attainment among non-white students in America, Singleton and Linton (2006, 64) argue that, 'in courageous conversation, the solution is revealed in the process of dialogue itself.' Drawing upon Freirean pedagogy and Critical Race Theory, this paper will look specifically at youth and community work students' responses to critical conversations on race. These conversations were intended to enable both black and white undergraduate and post-graduate students to reflect upon their own attitudes, experiences, practices and beliefs on issues of race and racism. The article will look at the contemporary and historical context of Moss Side, Manchester, UK before turning to consider whiteness, informal education and Critical Race Theory. It will then evaluate the course as a location for the development of conversations of protest. Whilst being critical the conclusion asserts the continuing need for courses of this type in the development of anti-racist practice and institutions.

'Our Moss Side'–historical and contemporary contexts

Point 5.1.4 of the National Youth Agency (NYA) Statement of Principles (2009, 8), is concerned with the promotion of justice for young people and in society in general. In practice this includes:

(1) Promoting just and fair behaviour, and challenging discriminatory actions and attitudes on the part of young people, colleagues and others.
(2) Encouraging young people to respect and value difference and diversity particularly in the context of a multicultural society.
(3) Draw attention to unjust policies and practices and actively seeking to change them.
(4) Promoting the participation of all young people and particularly those who have traditionally been discriminated against, in youth work, in public structures and in society.
(5) Encourage young people and others to work together collectively on issues of common concern (www.nya/../uploads/2014/06, accessed 10 August 2016).

Consistent with the NYA Statement of Principles,the 'Our Moss Side' documentary was one of the strategies which young people used to challenge what they regarded as the biased and discriminatory views of the Manchester Evening News headline, *Top Tory Compares Moss Side to the Wire*. Following his 2009 visit to the X calibre anti gun and gang unit at Greenheys police station in Moss Side, Chris Grayling the then Conservative Party Shadow Home Secretary, compared the streets of Britain to the violent American television programme *The Wire*. This comparison was based on what he claimed to have been his experience of urban warfare in Moss Side which has the largest African-Caribbean heritage people in Greater Manchester. When compared to the population of England and Wales, proportionally there are more people in Moss Side aged 0–34 but fewer residents over the age of 35 (Davis, Watt, and Packham 2012, 7). Moss Side is now home to a high number of young black people whose grandparents and in some cases their great grandparents were amongst early post-war migrants from the Caribbean. The Windrush Millennium Centre and the West Indian Sports and Social Club in Moss Side symbolises the community activism of early migrants such as the late Euston Christian. During the Second World War he was a member of the Jamaican Royal Air Force but was later invited to join the British Royal Air Force. At the end of his service he returned to England on the Empire Windrush which landed at Tilbury on the 22 June 1948. In their book in celebration of the 50th anniversary of the arrival of the Empire Windrush, the brothers Mike and Trevor Phillips (1998, 4) describe the events of that day as a ' journey through the gateway of history, on the other side of which was the end of Empire and a wholesale reassessment of what it meant to be British' (1994, 4). It is against this background that the making of this documentary was also an opportunity for young people to participate in conversations and to listen to stories on equality and diversity as in the experiences of their parents, community activists, Saturday supplementary school tutors as well as youth and community workers. Being involved in these conversations was important because:

> People will act on the issues on which they have strong feelings. There is a close link between emotion and the motivation to act. All education and development projects should start by identifying the issues which the local people speak about with excitement, hope, fear, anxiety or anger (Hope and Timmel 1984, 8).

The young people's conversations with members of the wider community included people such as myself who for a number of years lived and worked voluntarily in Moss Side. As an adult my voluntary activities have been primarily linked to the political, cultural and educational work of Abasindi Black Women's Co-operative and that of the Nia Cultural Centre for African and Caribbean performing arts. However, my involvement began with my parents who were amongst the first group of post-war migrants to purchase their house in Moss Side. During the 1960's black people's choice of where they lived was quite restricted. This was largely due to racial discrimination both in the allocation of council houses and access to mortgage facilities. Having joined the British Air Force out of what he described as a sense of patriotism during the Second World War, another of the founder member of the West Indian Sports and Social Club in Moss Side was amongst those who experienced racism in his choice of neighbourhood.

> I bought my first house and sold it after a year. I remember neighbours collecting a petition against my living in that road. I told them then, that my money paid for the house. The area was predominantly white but as time went by, the white owners sold to black families and moved out of the area. (Roots Oral History Project 1992, 16)

During the 1960's house prices also rose steadily and finance companies began to work out special short term loans to black people who could not afford the normal deposit. The unusually high interest and repayment rates, forced many of these families including my parents to let rooms to other black workers. Unlike the houses in Jamaica where I was born, in Moss side my parents occupied the two rooms on the ground floor. This included the 'front room' which was only used on special occasions and much of the time as a family was spent in the kitchen/dining room. The two rooms on the first floor were rented out by another family whilst the two girls in the family lived in the attic room and the cellar was used for storing coal. In some black households the cellar was also a space for socialising and what became popularly known as cellar parties. Although there were kitchens on both the ground and first floors, we only had one bathroom. This situation was not unusual in that it was quite common for two or three families to share a single kitchen and bathroom. Despite these difficulties, by 1969 the majority of black people living in Moss Side owned their own houses (Phillips 1974).

In 1968 Manchester City Council announced plans for demolishing and rehousing the entire area. The responses to this announcement included a highly publicised political campaign opposing the destruction of the area. This campaign was spearheaded by a local Housing Action Group and membership of the group included the late Kath Locke who has been described as one of Manchester's most 'powerful and feisty fighters for the people's rights' (Watt and Jones 2015, 41). However, faced with rapidly rising house prices in other areas and continued racial discrimination, the majority of black householders including other family members had no choice but to accept tenancy from the local authority, in that compensation payments were generally too low to provide an equivalent sized house elsewhere. This was the experience of the Roots Oral History Project participants. 'Our first home was in Meadow Street in Moss Side … I lost this house under a compulsory purchase order from the court. They paid less than what the house was worth and I subsequently made a loss on it'. Another family stated that they received no compensation, 'I bought my house a year after I came to England in 1955 for £500.00. The house was demolished by the government after 14 years and I was never paid a penny' (Roots Oral History Project 1992, 7–8). The demolition of their houses meant that black people were dispersed to areas such as Wythenshawe, the largest district in Manchester with an equally large housing estate which was created in the 1920's. This dispersal was short-lived because a significant number of black people refused to be re-located on housing estates away from the city centre. Night workers often found themselves with little or no access to public transportation. Furthermore, the local all white schools were also reluctant to accept black children from the inner city. Confronted with the daily realities of racism and prior to the development of the Alexander Park estate in Moss Side, the majority opted to be re-housed in the neighbouring district of Hulme (Phillips 1974). The experience of my parents and other early migrants represented what the Roots Oral History project (1992) described as a 'rude awakening'.

One of the women who had been living in Manchester since arriving from Guyana in 1961 was also a member of the North West Conciliation Committee. Following the 1965 Race Relations Act, herself and other members of this committee had responsibility for dealing with discrimination in public places. In conversation with her in 2010 she spoke about the fact that

> These were days of overt racism and members took it in turn to visit Public Houses where public notices were displayed in windows which read: 'No Irish, No Blacks, No Dogs'. Our duty was to inform landlords that these were now illegal and they must be removed and that the practice of restricting access had to cease.

Both Irish and African-Caribbean people were the recipients of crude anti-immigrant hostility by the English, for the latter group, Mirza (1992, 158) describes it as far more 'powerful, direct and insidious'. In her discussion on the opposition between Englishness and 'immigrants' as in the expectations of white neighbours and the perceived disruption by West Indians of private and public spaces, Huxley (1964, 43–44) wrote:

> … most West Indians like loud music, noise in general, conviviality, visiting each other, keeping late hours at weekends, dancing and jiving … Most English people prefer to keep themselves to themselves and guard their privacy. Ours is a land of the all, the high fence, the private hedge – all descendants of the moated grange.

Huxley was writing at a time when the Cypriots, Maltese and Italians were regarded as amongst those who led respectable family lives. Polish people eventually earned respect as conformist and solid householders who were keen to improve the state of their properties. On the other hand the people from Southern Ireland were 'as bad as the darkies' Patterson (1965, 179). Although, originating from amongst white elites as in the title of Patterson's book *Dark Strangers*, irrespective of class and status expressions of whiteness in its various forms, subsequently became acceptable to the white community (Jensen 2011). Taken for granted every day acts of racism were thus seen as the norm. From these early beginnings of Black communities racialised experience in Moss Side and its continuation into the contemporary period, we can see that any anti-racist informal education pursued has to include a Critical Race Theory perspective.

Whiteness, informal education and Critical Race Theory

As informal educators, youth and community workers are involved in activities aimed at facilitating learning and the social and personal development of young people and/or people in communities, to take collective action. This takes place in a range of settings and participation is voluntary. The NYA defines participation as the 'process by which children and young people influence decision-making which brings about change in them, others, their service and their communities' (Young 1999, 22). The characteristics of informal education draw heavily on the philosophy and practice of the Brazilian popular educator Paulo Freire. According to Cornel West, Freire has the distinctive talent of being a profound theorist who remains 'On the ground' and a passionate activist who 'gets us off the ground' (Ledwith 2005, 53). In Freirean pedagogy, dialogue is a form of critical conversation which, from a humanist perspective, engages people both intellectually and emotionally. It is the basis of praxis and the link between reflection and informed political action (Packham 2008). This means that conscientisation is the process of 'learning to perceive social, political and economic conditions and to take action against the oppressive elements of reality' (Freire 1972, 15). The use of experience and problem posing represents a rejection of the anti-conversation model of education. Freire (1972) viewed education as a political process which was either domesticating or liberating. He regarded the 'banking' approach to education as a form of social control in that it is a process which involves the 'depositing' of information by the powerful educator into the minds of the uncritical learner. In turn this information is regurgitated by the learner through various forms of assessment, the results of which will determine one's life chances as defined by Max Weber ([1948], 2009). Despite significant differences between learning within a formal context and that which occurs in the process of informal educational activities, Beck and Purcell (2010, 29) argue that

> Education whether formal or informal or community based is a potentially dangerous process. As youth and community workers, we bring our cultural values into the communities we work with. This is inevitable. The danger is if we do so without reflection or criticism we will unwittingly impose our culture thereby disempowering the very people we want to empower. Friere describes this practice as cultural invasion. The process which makes communities see realities through the eyes of the 'invaders', accept imposed norms and values, see themselves as inferiors and become powerless. This leads to a situation where people develop a Culture of Silence.

Racism leads to cultures of silence through it is taken for granted in daily practices. Indeed, Critical Race theorists Delgado and Stefancic (2001, 6–9) assert:

> Racism is ordinary … the usual way society does business, the common everyday experience of most people of colour in this country … Because racism advances the interest of both white elites (materially) and working-class people (physically), large segments of society have little incentive to eradicate it.

Critical Race Theory maintains that the source of 'whiteness' is that which is rooted in the process of oppression and domination (Knowles and Lander 2011). There is a recognition that structural racism can exist without people's conscious intention or awareness.

Paradoxically, discrimination can occur even when there is the intention of equality. However, power is held by those racialised as white. Whiteness is often seen as a neutral identity model whose members do not consciously or unconsciously see themselves as different. Therefore, it is difficult for them to assume knowledge and understanding of the oppression of 'other' races and ethnicities. Unlike their earlier experiences of persecution and extreme forms of discrimination, the re-classification of Irish and Jewish people means that they are able to access privileges associated with whiteness. Thus, it is argued that being classified white is dependent on how one is perceived by white people with power (Jensen 2011). In his reflections on whiteness and race privilege one of the youth and community work students on the course commented on the fact that:

> As a young male of Caribbean heritage, I am of a light complexion and I am often mistaken for a Caucasian male…. I despised the fact that my lighter complexion was accepted more than a darker complexion and subsequently this influenced my self-concept and led me to feel Caucasian.

Although this young man's first name is of West African origins, his complexion and the absence of visible features that are constructed as being 'common' amongst people of African descent rendered him 'White by proxy' (Knowles and Lander 2011, 58). Unlike his darker skinned peers, he was not penalised for not having European features. He was nevertheless aware of the stereotypical experiences of dark skinned young black men and the power of him having access to privileges on the basis of being regarded as white. This tendency towards white sameness denies differences in areas of race, racism and ethnicity (Knowles and Lander 2011). The denial of differences is often associated with colour blindness and the focus of training sessions on diversity is on ensuring that participants are not made to feel uncomfortable (Singleton and Linton 2006). In commenting on this denial of differences, Audre Lorde stated 'It is not our differences that divide us but our failure to recognise, accept and celebrate these difference' (https://www.goodread.com/author/quotes/18486/AudreLorde, accessed 30 June 2016). Failure to recognise, accept and celebrate point to the necessity to frame informal work with these issues in mind and that intersectional thinking is its bedrock.

Intersectional race equality and diversity and informal education

The students on the undergraduate and postgraduate youth and community work degree programme were amongst those enrolled on courses at one of the three well established universities in Greater Manchester. These courses seek to attract people from diverse communities who are keen to develop their skills in working with young people and adults in informal and community-based settings. The specialist units on conflict and creativity, gender and sexuality, race, faith and diversity are intended to address issues in relation to empowerment, transformation, equality and diversity. The extract below is an example of the ways in which both undergraduate and postgraduate students during the course of their work-based placements are able to develop their skills as critical reflective practitioners and informal educators.

> … there has (in the nearly 3 years that I have worked there) never been any overt and sustained discussion on engagement with other forms of difference based on ethnicity or race. Inequality has been largely based on class and opportunity and recently disability in the form of an autism engagement project … In the past few weeks I have begun to realise this is the situation and

question it … working in informal education I want to question (and probably know it on an unconscious level to be a reality), whether it is the same practice as formal education, even if on appearance there is more scope for creativity and challenging the status quo, subconsciously we are biased (post-graduate student).

During the period of my own practice placement as youth and community work student in a predominantly white working class area, my experiences of overt expressions of racism is that which influenced my decision to work with young people primarily within an education/community-based setting. Over the years this has included responsibility for co-ordinating a college-based mentor service. This service was developed specifically to support African-Caribbean and South Asian heritage students that were enrolled on courses at one of the local Further Education colleges. The mentor service also provided placement opportunities for youth and community work undergraduate students. The mentors included a high number of black men and women from a range of professional backgrounds. Despite the programme's positive impact especially the development of the 'Rites of Passage' programme for black boys which featured in an hour long television documentary, Majors (2001) argues that without changes at an institutional level, mentoring programmes cannot make a significant difference to the life chances of mentees.

As an informal educator, I was also involved in a range of community activism and on joining the youth and community work teaching team at the university, I was already familiar with the team's work in areas of equality and diversity. One of my areas of teaching included the module on equality and diversity which was aimed at enabling all students to develop a critical awareness of the significance of racism and ethnic differences. Although black and South Asian students are consistently represented on the youth and community work programme, with the exception of one year group the majority of the students were white young people from a range of socioeconomic backgrounds. This majority presence was not reflected in the take-up of the equality and diversity module which was initially an elective. Apart from the odd white student, this elective was the only space in which black and South Asian students were always in the majority. For these students the focus of the module on issues of race and racism was not part of their experiences at school, particularly that which is based on what critical race theorists describe as a 'white supremacist masterscript' (Swartz 1992). Swartz (1992, 341) argues that:

> Master scripting silences multiple voices and perspectives, primarily legitimising dominant white, upper class, male voicings as the 'standard' knowledge students need to know. All other accounts and perspectives are omitted from the master script unless they can be disempowered through misrepresentation. Thus, content that does not reflect the dominant voice must be brought under control, and then reshaped before it can become a part of the masterscript.

As an elective the popularity of the equality and diversity module amongst black and Asian students was based on its 'respect for basic human and indiviudal rights, respect for difference, a commitment to empowerment and participatory democracy, collective action and voluntary (consenting) participation' (Packham 2008, 14). The module provided a 'safe' space for storytelling which is one of the three main goals of Critical Race Theory and that which is central to Paulo Freire's discussions on the relationship between transformation and critical consciousness. In the field of community development the use of stories encourages participation by way of listening and understanding. It is through the act of being listened to that the storyteller experiences a sense of belonging and increased levels of confidence. Storytelling is a form of communication that is less fearful, less radically associated, yet capable of transformation as it stimulates critical consciousness (Ledwith 2005, 13). The discussions on 'stop and search,' refugees and asylum seekers and the educational experiences of black British boys of African-Caribbean heritage were some of the topics which enabled students and lecturers to become engaged in what Batsleer (2008) describes as consciousness-raising conversations. It was an opportunity for black and South Asian students to gain their 'voice' and to enable aspects of the self that may usually be silenced to emerge. The students were encouraged to name their realities rather than simply adapt to the ways in which their realities are named for them (Batsleer 2008, 19).

Reflecting on their own experiences of the education system the majority of the students that were enrolled on this elective expressed concerns that the module with its focus on race was not a core element of the youth and community work programme. One student said

> My professional reason for choosing this subject started when I witnessed an incident that occurred in the first year of university, It was an occasion when a young black male entered the youth club where I was carrying out my placement and was very agitated and angry. He was shouting 'I did nothing wrong. I was on my way here. Nothing, absolutely nothing. I am f…king tired of the police doing this to me.' Immediately this led me back to my childhood, how were the same words still being shouted by a black male? From this point my interest surrounding the issue became so intense, I began to read many books, theories and newspaper articles that addressed the issue.

The move from an elective to a core module was intended to enable all students to examine their own attitudes and beliefs in relation to racial and ethnic differences. It was also an opportunity for students to develop their skills, knowledge and understanding of anti-oppressive and anti-discriminatory practices. During one of the sessions on myths and realities as in the experiences of refugees and asylum seekers, some of the black female students were in a visible state of distress. This session was facilitated by a postgraduate student whose experiences as a refugee were not unlike that of the student who had twice experienced detention. It was at this stage that the majority of students realised that for the last two years they had been studying alongside people who bore no relationship to the media's negative portrayal of refugees and asylum seekers. Instead they were seen as victims of government policies aimed at disempowering, controlling and humiliating them. On the issue of empowerment for one of the students it meant 'getting the necessary information about my rights and being treated like a human being'. Another student stated 'it is to be able to speak out your problems and get your voice heard'. Based on these experiences, Muhawenimana (2016) argues that youth and community workers should regard research in these areas as a starting point for enabling asylum seekers to reflect on their circumstances and to take collective action for change. As a group, the students' sensitive response to these personal stories of discriminatory and oppressive practices and their choice of essay topics further support the view that courageous conversations on race also involve a recognition that:

> The academy (institution) is not a paradise, but learning is a place where paradise can be created. The classroom with all its limitations, remain a location of possibility. In that field of possibility we collectively imagine ways to move beyond boundaries, to transgress. This is education as the practice of freedom (Hooks 1994, 20).

From a youth and community work perspective, the majority of students found this session extremely informative but the fear of engaging in conversations on race and racism was clearly evident amongst some of the white students who chose not to attend the sessions but did not openly state their reasons for doing so. This was with the exception of the white male student who after attending two sessions had decided that the module was irrelevant because it was all about race. In voicing his objections, the student went on to explain that some of his friends were black so race for him was not an issue but he was concerned that nothing had been said about the attack on two young people for being gay. The student was present when the group was informed that race and sexuality were included in the key themes to be discussed. The decision to include race and sexuality was in part influenced by one of the youth and community work students' comments on the experiences of a black gay man who was refused entry to one of the clubs in Manchester's gay village . The reasons for his refusal were couched in racial stereotypes in that he was described as being, 'Black, Big and Bad'. Without commenting on the issue of race the student responded by giving examples of 'straight' people who had been refused access. Given his concerns about issues in relation to sexuality, it was somewhat disappointing that he had failed to attend the session on *Race and Sexuality* which was facilitated by a black male psychologist who is also gay. The attitude of this student and that of those who remained silent could be attributed to a refusal on their part to accept the existence of racism or a failure to recognise racism in an institutional context, and the significance of race in their personal and professional interaction. The response of students in England to conversations on race is not unlike the experience of Taylor, Gillborn, and Ladson-Billings (2009, 17) in the USA, who stated that, 'Outside the supportive confines of our own institution, we were met with not only the expected intellectual challenges, but also outright hostility. Why were we focusing only on race? What about gender? Why not class? Are you abandoning multicultural perspectives?'

The students in this unusually black majority cohort had also failed to acknowledge that in planning the module there was space for them to do presentations on areas of inequality and difference that were of personal and political interest to themselves. These student-led sessions included presentations on *the Lesbian and Bisexual Women's Community and Celebration of Differences and Acceptance of the LGBT community in Cyprus*. In citing Human Rights Watch, The presentation concluded by stating that, 'Sexual orientation and gender identity are integral aspects of ourselves and should never lead to discrimination or abuse.' The discussion on these themes was not reflected in the students written feedback instead the white students wrote that contrary to its title, the module was not equal in that the focus on race meant that there was no opportunity to challenge other forms of discrimination and that the group presentations were not sufficient a space in which to do so. The presentations by black and South Asian students included the celebration of Diwali *in Bristol, Diverse Culture and Transition in Moss Side and the Motherland – Africa*. As opposed to anti-racism, multiculturalism was the focus of these presentations but it was indeed an opportunity for all students to have their voices heard. This is significant because:

> Students from marginalised groups enter the classroom within institutions where their voices have been neither heard nor welcomed … to hear each other is an exercise in recognition. It also ensures that no student remains invisible in the classroom. (Hooks 1994, 41, 84)

Conversations of protest: hearing, recognition, visibility?

The report by Hick et al. (2011, 12, 15, 17) *on Promoting Cohesion Challenging Expectations*, draws attention to the fact that, 'there is ample evidence that one-off modules on any topic do not suffice to make lasting behavioural change'. This report is based on the findings of a research team from the University of Edinburgh and Manchester Metropolitan University.

Interviews to find out how both universities were dealing with race equality issues were undertaken with 31 lecturers that were involved with primary and secondary teacher education in Scotland and England. The reluctance to engage in critical conversations on race was also prevalent amongst trainee teachers and according to one of the female lecturers, 'I find that generally – two things that they don't like talking about are race and class. Race definitely, class some people will talk about but they don't talk about race'. One of the male lecturers described the issue of race as 'thorny' and one that people feel 'I won't talk about it cause I don't want to be racist'. This was clearly evident in the response of interviewees to the theme on building an awareness of race equality issues.

> What I've found when I've talked with other staff who are from non-minority ethnic background on 'race' and ethnicity is that everybody is on the defensive. People are on the defensive often because they think you are suggesting that they are not working sensitively with students or (if) they are not aware of the issues they are racist (female secondary).

Gillborn (2008, 3) argues that

> racism is such a harsh word that some people feel uneasy about using it … the term is so forceful that most people react very defensively against any suggestion that they might possibly be involved in actions or processes that could conceivably be termed as racist.

Whilst conversation is one of the most important methods of engagement, it is evident that for both formal and informal educators, conversations on discrimination and oppression can prove to be both an emotive and threatening subject matter. It is therefore not surprising that some people may dig their heels in about particular issues (Thompson 1998, 147). The reluctance of white students cannot be attributed to the failure of the black facilitators to engage with them in a manner that was not conducive to learning. With the exception of the sessions on youth work from an anti-racist perspective and the Anthony Walker Foundation, the other sessions were facilitated by highly skilled black people from a range of professional backgrounds. These facilitators were themselves faced with some of the questions outlined in Taylor's (2005, 25) paper on *Is it All Black and White*?:

> I have been researching the Black/White dynamic for some years now and attempts to educate white community and youth workers in these areas have not been easy. Continuing issues for me as a Black lecturer/trainer are: will white students or professionals begin to discuss sensitive Issues around race. Will my Black identity be a barrier to committing to dialogue around race. And what might be the issues and attitudes that white workers may have towards Black people from a personal and professional perspective?

Much consideration was given to the personal and professional relevance of the topic areas for discussion as in the session on *Youth Work from an Anti-Racist Perspective*. This session was facilitated by a white female youth worker who has been very proactive in her conversations and practice on issues or race and racism. For the majority of white students this was one of their favourite sessions hence comments such as 'fantastic – it gave valuable skills for workshops and challenging discriminatory practices and to tackle issues in regards to racism with young people', The other favourite session was the one on the work of the Anthony Walker Foundation which was facilitated by a white postgraduate student. This session was described as 'really dynamic, upbeat and motivating', 'excellent presentation which made me aware of racism', 'really liked the games and enthusiastic vibe'. Throughout the session the focus was on hate crime as opposed to racism. The approach of the two white female facilitators undoubtedly minimalised the students' fear of being accused of racism.

The session on *Why it is important for us to explore differences and its implication for work in areas of community development and community cohesion* was described by some students as 'boring' and other students viewed it as an important topic area. The comments on *Race Equality in the Work place* covered a range of positive and negative comments from 'boring – no activity just talk, talk, talk', to 'really enjoyed this session, good for policies relating to me and how to apply these in youth work'. In their comments on the session on *Stop and Search,* some of the students stated that the speaker's approach was 'engaging and interesting, I loved it, it was amazing, excellent', 'the police officer made me more comfortable and aware of human rights', 'could have been more interactive and this was all about black people and other ethnic groups they left aside'. The opportunity for students to talk about the experiences of other groups was present throughout the session.

Unlike the sessions on *Race Equality in the Workplace and Stop and Search* the session on *Raced identities in youth work*, was facilitated by an experienced youth and community worker. Yet, it was described by some students as 'really slow/quite boring', 'not engaging'. The session on *Celebrating Black History month* was described as their least favourite. On reflection one of the students stated that it was a 'provoking lesson and initially I was angry but after I spoke to others I realised that was the point.' Some students found it inform-ative, interesting and others a bit boring 'because the woman just talked without doing any activities'. One of the British South Asian students spoke about the ways in which the module had enabled her to challenge racism within a multicultural college environment. The assignments were also cited as a source of learning, to quote one student:

> I really enjoyed the essay as it threw up new theories such as Critical Race Theory which I had not heard of before. From this essay I gained more of an insight into institutional racism which is relevant to me as a black male in the UK and my understanding of how society treats me.

The repeated expressions of boredom amongst white students and their resistance to engaging in critical and courageous conversations on race is not altogether surprising. John (2006, 213) stated that, despite 'two full scale reviews of the National Curriculum, schools are still required to deliver a curriculum that is essentially geared for life in a mono-racial and mono-cultural environment'. When presented with the Black History Month Wordsearch, activity on famous black people, the majority of students age 18–27, had not heard of Mary Seacole, Steve Biko, Malcolm X, Shirley Bassey, Maya Angelou, and Marcus Garvey. Most had heard of Bob Marley but knew nothing about his country of origin. None of the students had heard of John Sentamu who was born in Uganda and is the Archbishop of York. In the exercise on popular stereotypes, the white students associated power with being white and poverty as a reflection of black people's socioeconomic position. The students were totally oblivious of the fact that Gatwick airport is owned by Adebayo Ogunlesi, one of Nigeria's leading businessmen (thenubiantimes.com/gatwick-airport-is-black-owned, accessed 24 November 2016, Nubian Times, October/November 2015). The college and university students who participated in the 2009 National Union of Students' Black Students research project also pointed to the need for a more diverse perspective especially in areas of history, arts and politics:

> Professionals, Black and white must contend with how they feel about themselves and their experiences to date … It is clear from my experience that many professionals only begin to consider how racism can be tackled when confronted with issues around race in a setting that is amenable. If we hold attitudes that encompass fear and ignorance then we will not get to the root cause of those attitudes and begin to address them. Professionals, especially within the youth and community work context need to be conscious individuals who want to create change for others. To do this they also need to make changes within themselves and not hide behind the powerful aspects of silence, fear and ignorance especially if they truly believe in equality. (Taylor 2005, 28)

The report on the Black Perspective in Community and Youth Work conference in 2005, makes reference to Arshad's article on *Anti-racist community work*. In this article Arshad argued that

> Failure to engage with understanding the processes and dynamics of racism from its roots in the 17th century to its manifestations as we approach the millennium will lead to continued confused practice on the type of strategies that might be adopted to ensure radical social change.

Without this form of engagement, modules on equality and diversity will continue to be regarded by white youth and community work students as that which they are at liberty to either accept or reject.

Conclusion

In the case of the youth and community work students and as argued in this paper it is evident that the shift from an elective to a core subject was for many students an unpopular decision. Based on the examples cited, it is apparent that for various reasons the majority of white students were uncomfortable with this decision as seen in their consistent efforts to undermine and trivialise debates on the topic area. This proved an effective strategy for silencing white students who were keen on participating. Black students also found themselves at a disadvantage and some felt that active participation in the discussion would adversely affect their relationship with the wider student group. There were those who dealt with this dilemma by not attending on a regular basis and for others the assignments were ways of expressing their views without fear of 'isolation.' Another disadvantage faced by black and South Asian students was the fact that they no longer had a space in which to discuss their experiences of race and racism. Instead, they now had to defend or deny these experiences. Despite these challenges, modules on equality and diversity are nevertheless important spaces for informal educators to engage in critical and challenging conversations on race.

Disclosure statement

No potential conflict of interest was reported by the author.

References

Arshad, R. 2005. *A Black Perspective in Community and Youth Work 2001–2004*. Conference Reports. University of Manchester.

Batsleer, J. R. 2008. *Informal Learning in Youth Work*. London: Sage.

Beck, D., and R. Purcell. 2010. *Popular Education Practice for Youth and Community Development Work*. Exeter: Learning Matters.

Davis, E., D. Watt, and C. Packham. 2012. *Aspiration and Engagement Strategies for Working with Young Black Men*. Manchester: Community Audit & Evaluation Center, Manchester Metropolitan University.

Delgado, R., and J. Stefancic. 2001. *Critical Race Theory: An Introduction*. New York: New York University Press.

Freire, P. 1972. *Pedagogy of the Oppressed*. London: Penguin Books.

Gillborn, D. 2008 *Racism and Education: Coincidence or Conspiracy*. Oxon: Routledge.

Hick, P., R. Arshad, L. Mitchell, D. Watt, and R. Roberts. 2011. *Research Report – Promoting Cohesion, Challenging Expectations*. Manchester: Manchester Metropolitan University.

Hooks, B. 1994. *Teaching to Transgress: Education as the Practice of Freedom*. London: Routledge.

Hope, A., and S. Timmel. 1984. *Training for Transformation: A Handbook for Community Workers (Book 4)*. London: ITDG Publishing.

Huxley, E. 1964. *Back Street New Worlds: A Look at Immigrants in Britain*. London: Chatto & Windus.

Jensen, R. 2011. "Whiteness in Caliendo." In *The Routledge Companion to Race and Ethnicity*, edited by S. M. Caliendo and C. D. Mcllwain, 21–28. London: Routledge.

John, G. 2006. *Taking a Stand*. Manchester: Gus John Partnership Ltd.

Knowles, G., and V. Lander. 2011. *Diversity, Equality and Achievement in Education*. London: Sage.

Ledwith, M. 2005. *Community Development, A Critical Approach*. Bristol: The Policy Press.

Majors, R. 2001. *Educating Our Black Children: New Directions and Radical Approaches*. London: Routledge.

Mirza, H. S. 1992. *Young Female and Black*. London: Routledge.

Muhawenimana, V. 2016. "Protection or Continued Persecution: The Dilemma Faced by Youth and Community Workers Working with People Seeking Asylum in the United Kingdom." Unpublished diss., Manchester Metropolitan University.

Packham, C. 2008. *Active Citizenship and Community Learning*. Exeter: Learning Matters Ltd.

Patterson, S. 1965. *Dark Strangers – A Study of West Indians in London*. London: Penguin Books.

Phillips, R. 1974. *The Black Masses and the Political Economy of Manchester*. London: The Black Liberator.

Phillips, M., and T. Phillips. 1998. *Windrush: The Irresistible Rise of Multi-racial Britain*. New York: Harper Collins.

Roots Oral History. 1992. *Rude Awakening, African/Caribbean Settlers in Manchester – An Account*. Manchester, NH: Roots Oral History Project.

Singleton, G. E., and C. Linton. 2006. *Courageous Conversations About Race - A Field Guide for Achieving Equity in Schools*. Thousand Oaks, CA: Corwin Press.

Soni, S. 2011. *Working with Diversity in Youth and Community Work*. Exeter: Learning Matters.

Swartz, E. 1992. "Emancipatory Narratives: Rewriting the Master Script in the School Curriculum." *Journal of Negro Education* 61: 341–355.

Taylor, A. 2005. *Conference Report, A Black Perspective in Community and Youth Work*. New York: Manchester University.

Taylor, E., D. Gillborn, and G. Ladson-Billings. 2009. *Foundations of Critical Race Theory in Education*. New York: Routledge.

The Nubian Times. 2015. *Gatwick Airport is 'Black-Owned'*. Accessed November 24 2016. thenubiantimes.com/gatwick-airport-is-black-owned.

The Quality Assurance Agency for Higher Education. 2009. *Youth and Community Work Subject Bench Mark Statement*. Accessed June 20, 2016. http://www.qaa.ac.uk/Publications/Documents/Subject-benchmark-statement-youth

Thompson, N. 1998. *Promoting Equality: Challenging Discrimination and Oppression in the Human Service*. London: Macmillan.

Watt, D., and A. D. Jones. 2015. *Catching Hell and Doing Well: Black Women in the UK – The Abasindi C-Operative*. London: Institute of Education, Trentham Book.

Weber, M. (1948) 2009. "Class, Status, Party." In *From Max Weber*, edited by H. Gerth and C. W. Mills, 180–195. London: Routledge.

Young, K. 1999. *The Art of Youth Work*. Dorset: Russell House Publishing.

From Liverpool to New York City: behind the veil of a Black British male scholar inside higher education

Mark Christian

ABSTRACT

This paper speaks to the experience of Black British male experience in academia, with emphasis on UK and US colleges and universities. It is an autoethnographic study in terms of relating, witnessing, and noting both learning and teaching experiences within the confines of higher education in the UK and US. The idea behind this paper is to highlight the need for greater access and opportunity for Black scholars to teach and study without stress and strain on their minds and bodies. There are a number of studies that confirm the efficacy in sharing one's experience as an 'insider' within academia as a person of color. But there are few studies that speak directly to the reality of teaching and researching within the context of Africana or Black Studies in higher education. It is noted that academia should be a place whereby liberal arts of all genres are accepted and respected; in this sense, there is still a long way to go before we can attest to the affirmative of this point of view.

Introduction

> Let justice roll down like the waters and righteousness like a mighty stream. Dr. Martin Luther King Jr. (Cited in Washington 1986, 659)

The Black British male scholar confronting isolation and the glass ceiling in higher education settings appears to be a commonplace reality, given the proliferation of studies. Yet, it is difficult to actually describe or articulate the 'personal experience' when one is both employed and bound to the foibles of vindictiveness and power games within a given college or university setting, where the notion of 'academic freedom of expression' is not upheld as well as it should be. Moreover, it is contended here that if one 'stands' philosophically in the academic area of Africana/Black Studies, the complexities of isolation and discrimination are more intense for the Person of Color. Overall, there is something tangible about being a 'Black Male Professor' that induces both fear and insecurity in many White counterparts, both male and female (Christian 2012; Jones 2000).

Let me begin with a caveat for the reader: this is not a victim's tale. I am in every sense a successful tenured/Full-Professor living and working in New York City. Throughout my academic career, I have received awards for my teaching ability, published consistently in scholarly journals, in mainstream academic venues, and/or with independent progressive and specialized outlets related to my fields. Moreover, I have actively participated in professional organizations; I currently sit on editorial boards; I am a manuscript reviewer for top scholarly journals in my field; I have organized and chaired panels at major conferences, presented papers at regional, national, and international conferences on a regular basis, published books, edited special issue scholarly journals, contributed book chapters and many encyclopedic essays, and completed a copious number of book reviews. I am a Senior Fulbright Scholar, and have been a Research Fellow at the Institute of Commonwealth Studies, University of London, and have made many service contributions to my profession and current employer. Indeed, presently I am the Chair in the Department of Africana Studies at Lehman College, CUNY. In a real sense, I am an achiever in spite of the institutionalized racism in the United Kingdom (where I was born and raised) and in the US university systems (Baszile 2003, 2006; Chester, Lewis, and Crowfoot 2005; Christian 2012; Feagin 2002; Jackson and Johnson 2011; Law, Phillips, and Turney 2004; Stanley 2006). I have had over 25 years of higher education experience in the United Kingdom and the United States as an undergraduate, graduate student, grass-roots community educator, and university and college professor.

There has been a surge in studies relating to the experiences of Faculty of Color in higher education in the United States (Christian 2012; Jackson and Johnson 2011). This may be due to the fact that there is not much impact coming from diversity policies which effectively deny the prominence of 'race' as a major factor in unintended and disconscious racism. The term 'racism' is particularly complex as it relates to academia. Too often, racism is simplistically defined as a belief in superior/inferior human beings and using power to enhance this in various institutional practices. This may well be an outcome of racism but most often in higher education, there is the belief that educated people are bigger than the ignorance of racist speech and action. It is akin to the Ostrich putting its head in the sand and ignoring the surrounding danger.

We have to be brave and bold and we must confront the subtleties of racist outcomes. For example, in higher education, we can start by monitoring vigorously the lack of Black male academics that actually teach and work from the perspective of empowering their communities because there are so many Black communities. Too often, the Black male academic who works in academic circles is marginalized, undermined, and made to feel that he is not wanted (Christian 2012). To suggest that this statement is overblown or overly sensitive is folly because Black male academics that teach and conduct research to empower the social and cultural experiences of People of Color tend to suffer more problems gaining full-time employment, promotion, and long-term security in higher education (Christian 2012).

There is an acknowledgment to the work of White writers unpacking White privilege and White dominance (cf Feagin 2002, 2014; Ferber 2003) but more has to be done to ensure that 'race' or critical 'race' theory does not die within the matrix of diversity initiatives in higher education. Intersectionality is important too, as in the combining of Africana/Black Studies, but this focus should not diminish the importance of covert and overt assaults. For example, largely unacknowledged openly, Black male academics that are heterosexual and proud of their Black heritage, men who want to teach the historical and contemporary aspects of black culture find themselves often excluded from the 'center' in higher education if 'overly masculine.' In this sense, my racialized heritage is: Jamaican father, British mother of Spanish heritage, and I regard myself as Black British, of African-Caribbean heritage. Depending on one's maleness, it can be tricky negotiating one's body and mind in the confines of academia. Being 'too assertive' is not going to go down well with one's White colleagues. Therefore, most Black males in academia have to deal with various forms of emasculation, whether they admit it openly or not. To be 'overly Black and masculine' is dangerous in higher education, as it often is in the UK and US.

In terms of my current position, it is somewhat incongruous to embark on an arduously tricky task in which one shares one's innermost thoughts in relation to being a Black male scholar at a City University of New York (CUNY) setting in the Bronx. This is a college that has close to a 95% Students of Color population, made up of Dominicans, Puerto Ricans, as well as Africans from the continent, Caribbean, South and Central America, to name the key regions where the origins of students are found. Yet, at the same college, a 85–90% White faculty cohort exists. To discuss this anomaly is both a cathartic and courageous undertaking – if one is to be honest with the endeavor.

It is not as if the theme of Black faculty experience in higher education is novel; indeed, there are a number of studies that cover the issue (Benjamin 1997; Chester, Lewis, and Crowfoot 2005; Christian 2012; Feagin 2002; Jackson and Johnson 2011; Law, Phillips, and Turney 2004; Sotello and Myers 2000; Stanley 2006). In this sense, there is something tangible in the topic, something real, and it is problematic to talk about one person's experience as being merely anecdotal. Institutional racism exists and it manifests itself in higher education maybe not in overt ways, but certainly there is enough evidence to state there is much to be done in the United Kingdom or United States (my focus) before it becomes a remnant of a bygone age (Feagin 2002, 2014; Fitzgerald 2015). Ferber (2003, 328) contends that the situation confronting Faculty of Color working in such milieus is pretty grim:

> …many college campuses ignore issues of racism and inequality, or make only few, disparate efforts to address them. People of Color and women often face a 'chilly climate' on campus, and harassment and subtle discrimination are prevalent. What the sociological research tells us is that the problem is not an individual one; it is not as simple as weeding out a handful of prejudiced individuals who are poisoning the environment. Instead it is the environment itself that is the problem and must be changed….

No doubt the individual experiences faced by Faculty of Color are not entirely uniform, nor devoid of complex difference, yet there seems to be compelling evidence that institutional racism at the higher education level is palpable and sustaining itself as the second decade of the twenty-first decade is nearly to a close (Christian 2012; Feagin 2002; Jackson and Johnson 2011). Even with President Barack Obama, the acknowledged first African-American to hold this office, at the helm of the most powerful nation on Earth, institutional racism is manifest within US higher education (Feagin 2014, 194–200).

Intellectual background and development in the UK and US

Within the context of my experience as both a student and faculty Person of Color, who has studied and taught in both the UK and the US, Ferber (2003), 328 is correct when she views racism in higher education as an environmental issue. Moreover, the 'chilly climate' she speaks of is what I have endured on both sides of the Atlantic. However, it was difficult both as a student and faculty in the United Kingdom compared to having a wonderful graduate student and postdoctoral experience in the United States. I have had two full-time positions in the United States. I gained all my promotions at Miami University of Ohio from 2000 to 2011; from Assistant Professor through to tenured Full-Professor. Since August 2011, I have been based at Lehman College in the City University of New York system. At Miami University of Ohio, I encountered vindictive White administrators, and at my present position too. But I have also encountered good colleagues from White ethnic backgrounds too, so this is not a White-bashing ethnic perspective. It is simply a reflection on my experiences in higher education as a Black male academic.

Feagin (2014, 200) explains the tricky situation professionals of color face where White power presides:

> … A former top administrator at Harvard commented in [a] recent interview that in college settings professionals of color often have to signal to powerful Whites there that they are 'dominant-culture-friendly' and, if they are hired or promoted, that they will not make 'life miserable coming in here' and diversifying the organizational setting too much with people who are not White.

Again, in consideration of a 'chilly climate' for professionals of color in mainstream colleges and universities, the United Kingdom is more akin to an arctic blizzard, whereby Faculty of Color are simply conspicuous by our absence. The coldness of the system is pretty dismal. I remember there only being one faculty Person of Color during my undergraduate studies in Liverpool, England, at Liverpool Hope University (then named the Liverpool Institute of Higher Education) in the late 1980s to my graduation in 1992.

The faculty member of the sociology department was an African woman from South Africa, who focused on health issues. She was a Christian who had married a White South African Pastor and fled to England. Apart from her, there was not one Person of Color on the faculty. There were about five students of African heritage out of a 2000 plus student body campus. It turns out that I would become the first Liverpool-born Black male to have graduated from Liverpool Hope University. This may give an insight into how difficult it was completing my undergraduate studies in the United Kingdom as a male Person of Color.

Between studies, most of my time was spent volunteering in Black grass-roots education in Liverpool, at The Charles Wootton Center, an institution specifically designed to offer education to young Black adults. I had been a former student there and was first introduced to Black History lessons, such as Marcus Garvey, Malcolm X, Angela Davis, The Black Panthers and the general Civil Rights Movement. My Black Studies teacher was the late Ron 'Babatunde' Phillips (brother of the renowned journalist/broadcaster Trevor Phillips). He was a major inspiration to me but few in the British education system know of him or my connection to this profound Pan-African intellectual. Indeed, without his input in my life in 1980/1981, I may not have written this essay. On reflection, I actually gained the qualifications to go to university in Liverpool at the Charles Wootton; I was equipped to withstand any form of Eurocentric curriculum due to having had such a good teacher in Mr. Phillips. His teaching, by the way, took place in Liverpool 8/Toxteth at the height of the 1981 urban uprisings. I was a student of history while participating in Black British history.

Yet, for all that grass-roots education, one can never be fully psychologically equipped to deal with the mental isolation, a lack of community spirit, and the insidious racism that abounds in British higher education. Regardless of the coldness, I graduated from Liverpool Hope University top of my class in Sociology (1st Class honors) and did extremely well with my other major: American Studies (2:1 honors). Due to the loneliness that comes with a Black working-class social background while striving in a White middle-class environment, my time was spent in the library, usually reading from the *Journal of Black Studies* archives, which I found fascinating and inspiring. I read from volume one in 1969 up to the then early 1990s; it was absorbing to read the articles that spoke to the times that they were published at the height of the Black Power movement in the US.

On reflection, it was an intelligent way to educate myself in Black history and culture. For this was not manifesting itself via the general curriculum. To be fair, there was a decent White liberal male sociologist, a Marxist who lived in the best part of Liverpool and pontificated on the working-class plight from within the warm confines of the Ivory Tower, a man who introduced me to Frantz Fanon (1925–1961). After reading Fanon's *Black Skin, White Masks* ([1986] 1952), I knew education was for me a way to 'decolonize' the mind. In short, during my undergraduate years, and previous Liverpool Black grass-roots activism in education, I developed a strategy for surviving in mainstream (read White) higher education. I would do the work that was given to me via the curriculum, and continue my personal education into Black historical issues, figures, and events. It was a two-pronged educational experience that I developed early on that actually helps me survive now as a Black scholar in White spaces. Strategies for coping in White space are as relevant now as they were in the 1980s, and regrettably there are many experiences in the US that still need to be heard and understood (Stanley 2006).

A recent English study has expertly put together the 'hidden history' of Black British intellectuals and education in Britain (Warmington 2014). It brings together a number of key scholars in British multicultural education history that made it easier as a young scholar to navigate the corridors of education in England. It is refreshing to know that this history of Black British scholarship is now being recognized. Most of the scholars mentioned by Warmington (2014) are familiar to my intellectual heritage. Some of these Black British intellectuals influenced the ideas that were later developed in my studies, if not directly certainly indirectly (see Christian 2001, 2005, 2010).

Following on with my UK experience, in 1994, I was the only Black British scholar in the cohort of PhD candidates in sociological studies at The University of Sheffield, UK. There was an African student from Sierra Leone, who had to cut short his studies due to an immigration issue. Up to that period, we shared some great times, seeking out 'Black life' in the city of Sheffield. And we managed to actually find a grass-roots activist scene in the city. This helped me against the isolation within the walls of the university. It was difficult to see my African friend leave before he had gained his degree; nevertheless, I soldiered on and graduated in late 1997 with a PhD as the only Person of Color in the sociological studies department at The University of Sheffield.

Yes, it was not the most optimal of graduate experiences. Indeed, Sheffield was a culture shock for me after having previously spent 1992–1993 with the Department of Black Studies at The Ohio State University (OSU), where the majority of my professors were of African heritage. I did not have a single Professor of Color as an instructor during my graduate experience in the UK. Fortunately, I managed to gain the support of my advisors, Lena Dominelli, who left after my first year, and Ankie Hoogvelt. They are pretty powerful women intellectuals who had faced their struggles in an intellectual environment that was clearly White male dominated. Yet, my progressive advisors, for all their support, could not replace the 'warmth' of being with intellectuals that had experience of Eurocentric cultural hegemony.

Therefore, one can gage from this scenario that there are differences depending on individual location and institutional setting. Maybe had there been more Faculty of Color in

Sheffield, along with students and staff, then the time spent there would have been less isolating and 'chilly' in the blizzard sense. Having had the graduate experience at OSU, it was difficult to readjust back to basically an 'all-White' intellectual climate. For example, at OSU, I had discussed, debated, and dissected the works of pioneering Black sociologists such as W.E.B. Du Bois (1868–1963), E. Franklin Frazier (1894–1962), and Charles S. Johnson (1893–1956). Therefore, I knew that my education in sociology went deeper than that of most of my peers in the United Kingdom as we were not exposed to the works of Black sociologists.

What can knowing something about the intellectual history of Black sociologists have to do with surviving in predominately White university space? It was a fundamental aspect of survival as a Black student to have part of your education fit with your cultural background. I cannot overestimate the usefulness in studying the lives and works of historical Black contributions to sociological research and discourse. Maybe it is meaningless to those privileged White students who have limitless White intellectual role models to delve into *within* the mainstream sociology textbooks. Indeed, intellectual 'Whiteness' is normalized and becomes a taken-for-granted issue.

Most often, Black graduates, unless specifically involved with Black Studies/Africana Studies departments, are not exposed to Black sociologists in connection with my field. This is the same for Black students in psychology, anthropology, history, geography, and other disciplines in the social sciences and humanities. This situation can carry on right through one's career to being a professor in the academy, whereby Faculty of Color have little grounding in the varied intellectual works that emanate from their different cultural backgrounds, unless they do the added work, as I had to do in finding and learning about the many contributions. This is what led me to completing the Master's degree in Black Studies; and I would have gone on to complete the PhD in the same field but did not have the wherewithal to remain in the United States beyond my graduate experience for the Master's degree at OSU. Instead, I returned to the UK to complete a PhD in sociology while focusing my thesis on an aspect of Black British identity and social history.

Global White Supremacy and the Black scholar

Too often, the experience of the Black scholar, and generally Faculty of Color, in predominately White universities is set aside as a specific issue, unrelated to the broader history of racism. Charles W. Mills (2008) has made a powerful argument that in fact, 'White Supremacy' has been a largely unacknowledged political system functioning on a global scale since the 1500s. Moreover, it is a system that has profoundly made what the world is today. He contends that there is a degree of historical amnesia surrounding the origins of the US as a society related to such a global system, as he states:

> The problem here is a combination of historical amnesia (a forgetting of the facts) and conceptual blindness (a failure to see the implications of those facts, even when remembered). We need to see the European expansion over the world in the modern period for what it is- not as voyages of 'discovery' but as missions of conquest… In effect, Europeans came to rule the world. (Mills 2008, 97)

Mills (2008) is writing largely from a US context. However, the historical reality of his perspective is corroborated by British scholars, from a British context (Christian 2002a, 2002b; Fryer 1984). Indeed, Peter Fryer, in his book *Staying Power: The History of Black People in*

Britain (1984), has a chapter devoted to the 'rise of English racism' and its impact since the seventeenth Century. For Fryer (1984, 190),

> The chapter is central to this book, because the racism whose rise it outlines has been central to the experience of black people in Britain for the past 200 years. Long after the material conditions that originally gave rise to racist ideology had disappeared, these dead ideas went on gripping the minds of the living. They led to various kinds of racist behaviour on the part of many White people in Britain, including White people in authority. The chapters that follow show since 1784 black people in Britain have asserted their humanity, dignity, and individuality in the teeth of racist beliefs and practices.

If we take into account the works of both Mills (2008) and Fryer (1984), it is apparent issues of racism, such as slavery, colonialism, sharecropping, segregation, and second-class citizenship, are not only historical but contemporary phenomena embedded within all institutions and individuals in both conscious and unconscious ways. For too long, we have asserted that liberal arts educational institutions are largely free from institutionalized racism. It may be the case that most have a very good equal opportunities policy displayed for all to see on respective websites, but when evidence of Faculty of Color, students and Staff of Color is taken into account, the practice of equality does not meet with the theory of equality. Today, global White Supremacy Mills (2008) may not manifest itself in harsh, brutal, terms; but it does have relevance for People of Color in universities in the UK and the US. Listening to our voices may be something that can convince administrators that we are not making up stories, and/or telling tales (Benjamin 1997; Sotello and Myers 2000; Stanley 2006).

Africana/Black Studies in mainstream colleges and universities

It is a struggle for Faculty of Color to gain access to the mainstream in universities in the UK and the US, so what then for those that occupy positions in Africana or Black Studies departments and programs? This is an area that I am intimately familiar with having been most often on 'both sides of the fence.' To put it another way, I have been educated and I have been employed in the so-called traditional disciplines, along with being a student and advocate for Black Studies in the UK and US contexts (see Christian 2001, 2004, 2006a, 2006b, 2007). In short, more than 20 years inform my experience in the United Kingdom and in US contexts of Black Studies.

In that time, I have come to know that Black Studies has a contested position in the university, where it is akin to the unwanted stepchild (Christian 2006a, 700–701). There is little support for it from administrations, apart from lip service, in many universities; and too often those employed in the power positions have little interest in developing the field as a bona fide discipline. Nell Painter (2000) contended that Black professors in Black Studies were wrongly stereotyped as 'one group' and not given the credit that would come from the traditional departments and programs. This is a position that I cannot disagree with, I have both experienced and witnessed this at first-hand. If you make an attempt to address the administration, as I did on one occasion, then the cost can mean further marginalization. The fact remains that White Deans and Provosts are not all broad-minded and open to Africana/Black Studies, especially if one does not readily assume a 'dominant-culture-friendly' disposition (Feagin 2014).

In committing these thoughts to print, the voice inside me warns against expressing such opinions. However, it is the courageous side of my being that is winning through at this moment. However, with growing unemployment increasing at universities, a scarcity in resources, and a world that is wrongly escalating the notion and practice of color blindness (Bonilla-Silva 2003), it is little wonder that many Black scholars appear to lack courage in facing up to these common problems that we face in the workplace. The experience of vindictive White administrators in relation to Black faculty empowerment has been copiously noted (Christian 2012; Feagin 2014; Jackson and Johnson 2011; Jones 2000).

It is easier for some to walk away, or 'grin' through the pain, playing the game (read 'not challenging the administrations to support Africana/Black Studies'). Yet, the cost of the ticket to conformity is too high if one has a conscience. It is difficult for some to succumb and not take a stand, especially when it comes to the basic right and struggle for intellectual respect, academic freedom, and social justice. Having a right to promotion and salary increments that are based on individual productivity, not favoritism or other forms of subjective measure should be a fundamental reality in colleges and universities, regardless of one's racialized background or any other human attribute. For, at bottom, what we are all discussing in this journal is the varied experiences of discrimination faced by Faculty of Color in supposedly liberal institutions of higher learning. My contention here is that there is even greater struggle for Black male scholars connected to Africana/Black Studies, or other related programs, if the scholars are 'connected' to the field. Usurpers abound in Africana/Black Studies, often strategically placed by administrations to 'contain' the department or program by expounding the 'dominant-culture-friendly' paradigm. It has always been a case of 'making powerful White administrators comfortable' in a Faculty of Color experience (Feagin 2014, 200). If we comprehend the historical and contemporary manifestations of racism, then this notion and practice should be of no real surprise.

Fear and the White faculty administration

Much of the negativity in what happens to Faculty of Color is insidious; it is not openly direct racism these days. It is more of one cultural group containing the perceived radical element of Africana/Black Studies advocates. Fear plays a significant role in how White faculty interact particularly with male Faculty of Color. *Fear of the Black male* and the limited understanding of him abound in the Obama age and it is ubiquitous to American history and culture. The fear of Black male faculty (especially those in Africana/Black Studies) is something that has to be addressed openly as it is often swept under the rug by those promoting diversity initiatives that do little to improve the experiences of Black male academics.

Let us keep in mind that Faculty of Color are not the same but differ in many human ways; the same can be stated for White academics. That stated, it is also evident that 'White Privilege' operates as an invisible signifier for honor and prestige, giving the recipient a 'leg up' in most competitions for promotion and advancement in one's academic career (Feagin 2014; Ferber 2003).

According to Robert Jensen (Jensen 2003, 130), 'White Privilege' consists of a set of taken-for-granted entitlements that go largely unnoticed:

> What does it mean? [White Privilege]Perhaps most important, when I seek admission to a university, apply for a job, or hunt for an apartment, I don't look threatening. Almost all of the people evaluating me for those things look like me- they are White. They see in me a reflection of themselves- and in a racist world, that is an advantage. I smile. I am White. I am one of them. I am not dangerous. Even when I voice critical opinions, I am cut some slack. After all, I am White.

Jensen (2008) certainly makes plain that White privilege is tangible, yet something that is at the same time invisible to most Whites. The phrase 'I don't look threatening' jumps out to me, as it is at the heart of White–Black social interaction. For example, a good friend of mine at my present institution once stated out of the blue in relation to the students, 'I bet they're scared of you?' To be honest, I was taken aback. He really did not mean to be offensive. But it did get under my skin, for he could obviously see me as a 'threatening force' within our institution. As I stated earlier, this man is a good colleague of mine. We regularly go to lunch. Yet, 'race' as a topic is something that is very much avoided in our general conversations about life and the university. We can talk about aspects of our families, homes, lives of the mind in general, but not of the feelings that I have about racialized discrimination. Indeed, there is no one at my institution that I have yet been able to speak to about such matters.

Actually, I recall a Person of Color, but not of African heritage, once agreeing that I was clearly being discriminated against by an individual in power at our university. However, this Person of Color, who had a degree of power in the organization, stated in no uncertain terms that 'there is nothing I can do about it.' In other words, it was the way of the world here, and you just have to get on with it. It was dispiriting that someone with power felt that I was being discriminated against, but could not do anything for me as the person in question had more power. Therein lies the crux, it is about hierarchy in colleges and universities, power and privilege is conferred on those with power and privilege. Significantly, in higher education circles, the power typically lies within the hands of White administrators and faculty, and Faculty of Color are at the caprice of such people (Christian 2012; Feagin 2014).

Ironically, 'fear' is a two-way street. There is the age-old fear of the Black male, and then there is fear of and intimidation by White power among Faculty of Color. Fear of losing one's position; fear of those that are perceived as being able to do harm to one's career; fear of not advancing in the power structure of the college or university. Hence, Faculty of Color are not always likely allies for one another when one faces discriminatory practices, even if they have a degree of institutional power.

This is a crucial point, as too often we do not analyze the employment context of those that lack the courage to speak truth to power and who come from the less empowered Faculty of Color group. The individual that acknowledged I was being discriminated against was actually 'empowered' by the person discriminating against me. The issue of divide and conquer manifests itself when the White administrator uses a Person of Color who is 'dominant-culture-friendly' to put down the 'radical' or independent-minded Black male faculty member. It is a rather simple strategy that has been employed by White power advocates for centuries and continues to hold sway (Feagin 2014).

Life behind the veil

W.E.B. Du Bois often wrote about the experience of being 'behind the veil' as a Person of Color in White society. His work on double consciousness is central to this phenomenon, particularly in *The Souls of Black Folk* ([1982] 1903). In short, he argued that his people struggled to be both American and African-American in the US as the American side was in perpetual conflict, disturbing the development of the African-American. This analysis can be broadened to embrace the Person of Color surviving in a predominately White university setting.

All Faculty of Color, regardless of gender, have to develop a 'psychology of survival' while operating in predominately White institutions (Fitzgerald 2015). Some choose to grapple head-on with the subtle forms of insidious racism, while others put on the 'grinning mask' to deter assault on one's professional existence: with the strategy to 'live another day' in the university structure. One can empathize to some extent with those that choose to not confront the system in regard to perceived racist practices. We tend to always be out-numbered and out-maneuvered when it comes to having the support to make change for the better (Christian 2012).

Life behind the veil is complex in mainstream higher education. We who dwell in this experience are often conscious of its power in terms of developing a discernment about how to endure. Indeed, the racialized body does, after all, have to survive in these settings against the odds. Yet, even in the complexity of day-to-day survival, it is never uniform; it has both negative and positive realities, from desire to aversion. This then is the largely unacknowledged essence of what could be deemed the *internal-post-colonial experience* for Black male faculty. Yet, there is an irony. On the one hand, it is stated in college and university equal opportunity policies that one's racial identity, sexual orientation, class status, as well as one's physical, political, and other human attributes are not to be discriminated against. On the other hand, the racialized body is at the whim and caprice of conscious and unconscious racism (Bonilla-Silva 2003; Trepagnier 2010).

Leggon (2006, 216) provides an insight into the broader macro-statistical data that give 'meat' to what the realities are in terms of African-Americans' futile pursuit of faculty empowerment and visibility:

> When I received my Ph.D. in 1975, African-American faculty at PWCU's [Predominately White Colleges and Universities] were scarce, as I write this chapter 30 years later, African-American faculty are still scarce. Very little has changed insofar as the proportion of African-American faculty at PWCU's is about the same as it was in 1979–1980, women comprised 26% of all full-time faculty in higher education, African-Americans comprised 4%, and African American women only 2% of all faculty… In 1999–2000, of all full-time college and university faculty in the United States, women were 38%, African-Americans were 5%, and African-American women were 3%….

Leggon (2006) provides rather sobering statistical evidence of very little progress in regard to the 'embedding' of African-Americans and other Faculty of Color within mainstream higher education settings. Progress overall has been stagnant apart from that of African-American women, and this is rather slight. People of Color, African-Americans in this case, are conspicuous by their absence in many colleges and universities across North America. A token few are heralded as examples of success, but overall there are very few Black scholars actually doing progressive work.

However, I would contend that it is not just the racialized body, but one's intellectual endeavors that determine the level of discrimination one may or may not encounter. Too often, commentators speak merely of a lack of Faculty of Color in terms of 'physical diversity,' but it is important to have intellectual variance that brings different schools of thought to any given campus. Yet, this is not the case; most often, if a Black Scholar is 'too Black' in her/his writings, then often they will not be hired or promoted for a position to or within many colleges/universities. If one is hired, what is usually inevitable is the 'chilly climate' and undermining of one's intellectual contributions in terms of salary increments or promotion.

Stanley (2006, 11) maintains that what one researches as a Person of Color can be rewarded or unrewarded; she puts it this way:

> The question of what types of research are rewarded in the academy is answered with a lot of variance from faculty and administrators across academic disciplines. However, for Faculty of Color who reside in quantitatively oriented social science disciplines, research on race and racism can be differentially valued within the promotion and tenure system because of its political repercussions.

How one is rewarded is complex, individually based, and hard to analyze unless research is taken on one's peers in the social sciences, and on the measurements of, for example, starting salary, overall productivity, student and peer evaluations of teaching, consistency of publication record, and promotion opportunities (both won and lost). In other words, as hard as it is to measure who gets rewarded, and how, it also implies that it is uniformly uncomplicated for racialized discriminatory practices to occur and/or be camouflaged in routine evaluations of one's professional activities record.

The fact remains when I gained promotion to tenured Associate Professor in 2003/04, the majority of those evaluating my profile in the power structure of my university were White administrators and professors, from department level right through to the Board of Trustees. I got through due to meeting each criterion beyond the norm. Indeed, the Chair of Sociology at the time, and since, has stated that I crossed the promotion and tenure line 'by a mile.' On reflection, is not that what Faculty of Color have to be: better than the average to get employment, to get promotion? This is the common assumption that is relatively stable in the common sense reality of the social world. Black males in higher education are an endangered species because there is an ever-increasingly difficulty ladder to climb (Jackson and Johnson 2011).

Diversity Blues and the White liberal paternalism problematic

Diversity initiatives, for example, are something still very much 'in the making' and much work is to be done to improve multicultural perspectives on campuses across the US and the UK (Christian 2012; Feagin 2002, 2014). Interestingly, Faculty of Color in the social sciences and humanities will spend a great amount of time in the company of 'White liberal' academics. That is, those academics who consider themselves above the pettiness of racism, and even count themselves as allies to Faculty of Color. In my experience, overall, I would state that in most of the obstacles that I have confronted in academia, there has been a White liberal academic orchestrating the issue, not a radical or moderate right-wing conservative. Moreover, Faculty of Color are more likely to work closely with White liberals than any other group in the liberal arts fields.

Barbara Trepagnier (2010) puts forward an argument that 'silent racism' abounds in relation to well-meaning Whites and African-Americans. Again, this could possibly be extended to include Faculty of Color and White liberal academics. Having been around many White academics, that would be considered liberal in their outlooks on racialized discrimination, the Trepagnier's (2010) 'silent racism.' She defines the term thus:

> Two assumptions underpin the view that White people are either 'racist' or 'not-racist.' First, most Whites assume that racism is hateful; and second, most Whites believe that racism is a rare occurrence. These assumptions- that racism is hateful and rare- deny that racism today is often unintended and routine. Although blatant racism like that which occurred before the

civil rights movement occurs occasionally today, more often racism consists of routine acts of everyday racism that are not viewed as racist by the person performing them and therefore are not intentional. It is this unintentional racism… that produces a good deal of institutional racism and resulting racial inequality. (Trepagnier 2010, 3–4)

There is something tangible in what Trepagnier (2010) is arguing, but it is even more insidious than that. My interactions with White liberal academics have been varied. But one thing that seems to be commonplace among them in the academy is their 'fear' of a positive Black persona. In terms of positive, I am referring to a self-determined and independent approach to the life of the mind that is interested in the empowerment, culturally or otherwise, of peoples of African heritage.

Indeed, it seems that as long as the issue of 'race' is not mentioned in the presence of a White liberal academic, then things are going to be fine with regard to social interaction. However, once a topic on racialized discrimination is aired, there emanates an uncomfortable atmosphere whereby, for example, arms become closed tight into the chest of the White liberal, and eyebrows are raised. Unease permeates the conversation and it becomes clear that the only way out of this is to close down the conversation that caused the uneasiness. Therefore, nothing is achieved; we are in the same place, if not a worse one, than before the discussion began.

If talking to a White liberal academic about 'race' and racism is difficult, given their historical contours, inevitably, it is going to be harder expressing its reality to anyone else. Regardless of what the likes of Horowitz and Laksin (2009) want to argue, Persons of Color have an experience of racism in higher education. It is not imagined, most Faculty of Color tend to knuckle down and get on with things because it can exact great strain on one's life (Christian 2012, 131).

At bottom, it would be remiss to consider that the White liberal is always a 'friend' to People of Color (Trepagnier 2010). It really depends on just how much a Person of Color wants to live and act in a positive and independent manner or, on the contrary, with a victim mentality and posture employing a 'dominant-culture-friendly' character. In observing this social interaction, it seems that the more a victim perspective is portrayed, the more White liberal allies you will attract. More research is required on this phenomenon: White liberals and Faculty of Color in the academic sphere.

Conclusion

Edward Said (1994), 100–101 contended that it is a difficult task to speak truth to power. Frankly, this article has merely scratched the surface. More needs to be stated in relation to the proliferation of diversity initiatives that pay lip service to the issues of Black male faculty presence/absence in higher education. Nevertheless, it is a journey that continues to be travelled with the spirit of hope and a desire to improve the situation for the next generation of scholars at predominately White universities. When one considers the history of racism in the UK and the US, these societies have come some way, but certainly there is a long way to go before we can state: *all for one and one for all*. My biography-history has endured President Obama's rise and it is odd that I do not feel particularly empowered in my professional world.

In fairness to the US higher education scene, I owe this nation so much. It has given me a career that I enjoy and could not have earned in the UK, regardless of the racists that try to rain on my parade. I have made friends from many cultural groups across the human spectrum. But I am not naïve, there is room for improvement in the experience of those who teach and research in the area of Africana/Black Studies/Multicultural Studies in the US. I have witnessed discrimination in higher education against Faculty of Color, and I have suffered discrimination personally. However, I have yet to formally complain about it in the US because it is simply not worth the emotional hassle. Instead, if I am obstructed in the workplace by a disgruntled administrator, I write more, and I try to teach better. This is my positive response to those that may want me to falter, to fall, and to surrender. I will not succumb to those that want me to fail. Not without giving it my all in terms of work ethic. A strategy for survival is by tapping into the reservoir of historical personalities from past generations who endured far greater struggle than I could ever imagine.

Given the past, is there any wonder that institutional racism continues to live on as a 'changing same' in society, shifting its focus, moving with the times, but still dangerous? Most agree that there is no place for it, yet it continues to fester, even in academia. The legacy of racism continues to benefit the majority, and most do not even recognize it (Trepagnier 2010). Or if they are conscious of it, the propensity is to keep knowledge of their privilege silent. Those of us who 'live behind the veil' and progressive Whites need to continue to fight for social justice in all areas of society.

This article has provided a perspective primarily on the experience of one Black British male faculty in mainstream higher education. It is one area in the intersectionality of struggle that connects with the struggle of 'other' groups in society. Yet, the struggle of the Male of Color remains a salient feature in all areas of society, education being just one area. There 'fear of the Black male' is currently at crisis proportions within UK and US higher education and the broader societies. For example, the experience of Black males in the urban regions of the US and other nations such as the UK is something that needs to be highlighted and discussed more deeply in higher education. I am a mere witness to what can happen to even successful Males of Color in academia. White faculty and/or administrative 'fear' of Black intellectual independent-thinking males are something that needs critical attention in higher education – without fear or disdain.

Disclosure statement

No potential conflict of interest was reported by the author.

References

Baszile, D. T. 2003. "Who Does She Think She is? Growing up Nationalist and Ending up Teaching Race in White Space." *Journal of Curriculum Theorizing* 19 (3): 25–37.

Baszile, D. T. 2006. "In This Place Where I Don't Belong: Claiming the Ontoepistemological in-between." In *From Oppression to Grace: Women of Color Dealing with Issues in Academia*, edited by T. R. Berry and N. Mizelle, 195–208. New York: Stylus.

Benjamin, L., ed. 1997. *Black Women in the Academy: Promises and Perils*. Gainesville, FL: University of Florida.

Bonilla-Silva, E. 2003. *Racism Without Racists: Color-Blind Racism and the Persistence of Racial Inequality in the United States*. New York: Rowman & Littlefield.

Chester, M., A. Lewis, and J. Crowfoot, eds. 2005. *Challenging Racism in Higher Education: Promoting Justice*. Lanham, MD: Rowman & Littlefield.

Christian, M. 2001. "African Centered Knowledge: A British Perspective." *Western Journal of Black Studies* 25 (1): 12–20.

Christian, M., ed. 2002a. *Black Identity in the 20th Century: Expressions of the US and UK African Diaspora*. London: Hansib.

Christian, M. (2002b). "African Centered Perspective on White Supremacy." *Journal of Black Studies* 33 (2): 179–198.

Christian, M. 2004. "Unmasking the Mis-Education and Other Impediments to the Progressive Black Studies Scholar." *Africalogical Perspectives* 1 (1): 49–75.

Christian, M. 2005. "The Politics of Black Presence in Britain and Black Male Exclusion in the British Education System." *Journal of Black Studies* 35 (3) (January): 327–346.

Christian, M., ed. 2006a, May. The State of Black Studies in the Academy: Introduction to the Special Issue. *Journal of Black Studies* 36 (5): 643–645.

Christian, M. 2006b. "Philosophy and Practice for Black Studies: The Case of Researching White Supremacy." In *Handbook of Black Studies*, edited by M. K. Asante and M. Karenga, 76–88. Thousand Oaks, CA: Sage.

Christian, M. 2007, January. "Notes on Black Studies: Its Continuing Necessity in the Academy and Beyond." *Journal of Black Studies* 37 (3): 348–364.

Christian, M. 2010. "Black Studies in the UK and US: A Comparative Analysis." In *African American Studies*, edited by J. Davidson, 149–167. Edinburgh: University of Edinburgh.

Christian, M., ed. 2012. *Integrated but Unequal: Black Faculty in Predominately White Space*. Trenton, NJ: Africa World Press.

Du Bois, W.E.B. (1982) 1903. *The Souls of Black Folk*. New York: Signet.

Fanon, F. (1986) 1952. *Black Skin, White Masks*. London: Pluto; first published 1952.

Feagin, J. R. 2002. *The Continuing Significance of Racism: U.S. Colleges & Universities*. Washington, DC: American Council on Education.

Feagin, J. R. 2014. *Racist America: Roots, Current Realities, and Future Reparations*. 3rd ed. New York: Routledge.

Ferber, A. L. 2003. "Defending the Culture of Privilege." In *Privilege: A Reader*, edited by M. S. Kimmel, and A. L. Ferber, 319–329. Boulder, CO: Westview.

Fitzgerald, T. D. 2015. *Black Males and Racism: Improving the Schooling and Life of African Americans*. Boulder, CO: Paradigm.

Fryer, P. 1984. *Staying Power: The History of Black People in Britain*. London: Pluto.

Horowitz, D., and J. Laksin. 2009. *One-Party Classroom: How Radical Professors at America's Top Colleges Indoctrinate Students and Undermine Our Democracy*. New York: Crown Forum.

Jackson, S., and R. G. Johnson, III, eds. 2011. *The Black Professoriate: Negotiating a Habitable Space in the Academy*. New York: Peter Lang.

Jensen, R. 2003. "White Privilege Shapes the U.S." In *Privilege: A Reader*, edited by M. S. Kimmel, and A. L. Ferber, 79–82. Boulder, CO: Westview.

Jones, L., ed. 2000. *Brothers of the Academy: Up and Coming Black Scholars Earning Our Way in Higher Education*. Sterling, VA: Stylus.

Law, I., D. Phillips, and L. Turney, eds. 2004. *Institutional Racism in Higher Education*. Stoke on Trent, UK: Trentham.

Leggon, C. B. 2006. "Reflections from a Minority Faculty in a Majority White Institution." In *Faculty of Color: Teaching in Predominantly White Colleges and Universities*, edited by C. A. Stanley, 216–224. Boston, MA: Anker.

Mills, C. W. 2008. "Global White Supremacy." In *White Privilege: Essential Readings on the Other Side of Racism*, edited by P. S. Rothenberg, 97–104. New York: Worth.

Painter, N. I. 2000. "Black Studies, Black Professors, and the Struggle of Perception." *Chronicle of Higher Education*, December 15, pp. B7–B9.

Said, E. W. 1994. *Representations of the Intellectual*. New York: Vintage.

Sotello, C., and S. Myers. 2000. *Faculty of Color in Academe: Bittersweet Success*. Boston, MA: Allyn & Bacon.

Stanley, C. A., ed. 2006. *Faculty of Color: Teaching in Predominately White Universities*. Boston, MA: Anker.

Trepagnier, B. 2010. *Silent Racism: How Well-Meaning White People Perpetuate the Racial Divide.* 2nd ed. Boulder, CO: Paradigm.

Warmington, P. 2014. *Black British Intellectuals and Education: Multiculturalism's Hidden History.* London: Routledge.

Washington, J. M., ed. 1986. *A Testament of Hope: The Essential Writings and Speeches of Martin Luther King, Jr.* New York: Harper Collins.

Index

INDEX